Intro.

The Extracellular Matrix and Ground Regulation

The Extracellular Matrix and Ground Regulation
Basis for a Holistic Biological Medicine

Alfred Pischinger

EDITED BY Hartmut Heine
FOREWORD BY James L. Oschman
TRANSLATED BY Ingeborg Eibl

North Atlantic Books
Berkeley, CA

Copyright ©2007 by North Atlantic Books. All rights reserved. No portion of this book, except for brief review, may be reproduced, stored in a retrieval system, or transmitted in any form or by any means—electronic, mechanical, photocopying, recording, or otherwise—without the written permission of the publisher. For information contact North Atlantic Books.

Published by
North Atlantic Books
P.O. Box 12327
Berkeley, California 94712

Printed in the United States of America
Cover design by Gia Giasullo
Book design by Susan Quasha

The Extracellular Matrix and Ground Regulation: Basis for a Holistic Biological Medicine is sponsored by the Society for the Study of Native Arts and Sciences, a nonprofit educational corporation whose goals are to develop an educational and crosscultural perspective linking various scientific, social, and artistic fields; to nurture a holistic view of arts, sciences, humanities, and healing; and to publish and distribute literature on the relationship of mind, body, and nature.

North Atlantic Books' publications are available through most bookstores. For further information, call 800-337-2665 or visit our Web site at www.northatlanticbooks.com.

Substantial discounts on bulk quantities are available to corporations, professional associations, and other organizations. For details and discount information, contact our special sales department.

Original German edition: Das System der Grundregulation, 10/e
©2004 Karl F. Haug Verlag in MVS Medizinverlage Stuttgart GmbH & Co. KG, Germany

Library of Congress Cataloging-in-Publication Data
Pischinger, Alfred.
  [System der Grundregulation. English]
  The extracellular matrix and ground regulation : basis for a holistic biological medicine / Alfred Pischinger; edited by Hartmut Heine, et al; translated by Ingeborg Eibl; foreword by James L. Oschman.
       p.; cm.
  Includes bibliographical references.
  ISBN-13: 978-1-55643-688-8 (hardcover)
  ISBN-10: 1-55643-688-2 (hardcover)
  1. Extracellular matrix. 2. Physiology, Pathological. I. Heine, Hartmut. II. Title.
  [DNLM: 1. Extracellular Matrix—physiology. 2. Autonomic Nervous System—physiology. 3. Holistic Health. QU 350 P676s 2006a]
QP88.23.P5713 2006
611'.0182—dc22
                    2006034591

*In memory of my long-time medical
and
personal friend Franz Lutz, MD*
Baden, near Vienna

Contents

Foreword to the English Edition — xi
Introduction to the English Edition — xii
Foreword to the German Edition — xv
Introduction — xvi

PART ONE — Hartmut Heine, PhD, Doctor of Natural Sciences, Editor

1. **STRUCTURE AND FUNCTION OF THE EXTRACELLULAR MATRIX** — 3
1.1. Functional Unit of the Cell and the Extracellular Space — 4
1.1.1. Metabolism in the Matrix — 11
1.1.2. Chemical Sensitivity and Environmental Medicine — 16
1.2. The Extracellular Matrix as a Protein Regulator ("Slag Phenomenon") — 17
1.3. The Influence of Matrix Vesicles on Ground Regulation — 19

2. **LEUKOCYTOLYSIS** — 21
2.1. Regulation of the Tumor Extracellular Matrix — 30

3. **THE SIGNIFICANCE OF CHRONOBIOLOGY** — 32
3.1. Psychosomatic Stress Reaction Processes — 33
3.1.1. Controllable and Uncontrollable Stress — 34
3.1.2. Body Awareness and Stress Management — 35
3.1.3. Neurological Basis of the Stress-Reaction Process — 36

4. **TOPOGRAPHY OF EXTRACELLULAR MATRIX DISTRIBUTION** — 39

5. **STRUCTURAL COMPONENTS OF THE EXTRACELLULAR MATRIX** — 41
5.1. Glycosaminoglycans (GAGs) — 41

6. **SUGARS OF THE CELL SURFACE: THE GLYCOCALYX** — 43

7. **BASEMENT MEMBRANES** — 46

8. **PROTEOGLYCANS (PGs)** — 47
8.1. Synthesis of Proteoglycans — 49
8.2. Functional Aspects of CSPG-Hyaluronic Acid Complexes — 51
8.2.1. Dermatin Sulfate Proteoglycan — 52
8.2.2. Functional Aspects of Dermatin Sulfate Proteoglycan — 52
8.3. Heparan Sulfate Proteoglycan — 53
8.3.1. Functional Aspects — 53
8.4. Keratan Sulfate Proteoglycan — 53
8.4.1. Functional Aspects — 54

9. **STRUCTURAL GLYCOPROTEINS** — 55
9.1. Collagen Synthesis, Molecular and Supramolecular Structure — 55
9.1.1. Collagen Modification — 57
9.1.2. Functional Aspects — 58

9.2.	Elastin Synthesis, Molecular and Supramolecular Structure	59
9.2.1.	Functional Aspects	60
9.3.	Crosslinked Proteins	62
9.3.1.	Fibronectin	62
9.3.1.1.	Functional Aspects	63
9.3.2.	Laminin	64
9.3.3.	Chondronectin	65
10.	**ENERGY FLOW IN THE EXTRACELLULAR MATRIX**	66
11.	**IMMUNOLOGICAL BYSTANDER REACTION OF SUBSTANTIVE HOMEOPATHY**	70
12.	**SUGAR: EVIDENCE OF PRE-CELLULAR EVOLUTION?**	72
13.	**REFERENCES**	73

PART TWO—Otto Bergsmann, MD, Editor

1.	**GROUND SYSTEM, REGULATION AND REGULATORY DISTURBANCES IN A REHABILITATION PRACTICE**	81
1.1.	Physiological Regulatory Requirements	81
1.1.1.	The Organism: A Network System	81
1.1.2.	The Regulatory Cycle	82
1.1.2.1.	The Functional Elements	82
1.1.3.	Type of Regulation and its Quality, and Regulatory Disturbance	82
2.	**CHRONICITY AS BIOCYBERNETIC PROBLEM**	86
2.1.	Observations in Chronic Pulmonary Tuberculosis	86
2.1.1.	Pathogenic Investigation	86
2.1.2.	Therapeutic Results	87
2.1.2.1.	Breakdown of Hyper-reactive Allergy (Hyperergic) Reactions	87
2.1.2.2.	Increase in General Performance Capacity	87
2.1.2.3.	Treatment of Tension Syndromes and Tension Pain Syndromes with Neural Therapy	88
2.1.2.4.	Regulatory Therapy for Respiration and Circulation	89
2.2.	The Pathogenic Consequences	90
2.2.1.	The "Tip of the Iceberg"	90
2.2.1.1.	The Sensorimotor Control System	91
2.2.1.1.1.	The Segmental Regulatory Complex	91
2.2.1.1.2.	The Regulatory Control of the Musculature	94
2.2.1.1.3.	The System of the Muscular Maximum Point	96
2.2.1.1.4.	The Role of the Axial Structure in Regulatory Events	97
2.2.1.1.5.	Spinal Afferents and the Higher Order Control System	97
2.2.2.	The "Base of the Iceberg"	98
2.2.2.1.	The Ground (Matrix) System	98
2.2.2.2.	Information Transmission	99

2.2.3.	Regulatory Disintegration	101
2.2.4.	Minimal, Chronic, Persistent Stresses (Foci, Disturbance Fields)	101
3.	**DIAGNOSTIC PHENOMENA**	103
3.1.	Diagnostic Criteria	103
3.1.1.	Colloidal State	103
3.1.2.	Projection Symptoms of Internal Organs	104
3.1.3.	Acupuncture Points and Meridians	104
3.2.	Somatotopes	104
3.2.1.	Perfusion	104
3.2.2.	Humoral Parameters and Leukocytes	105
3.2.3.	Muscle Activity	105
4.	**THE POINT: THE WINDOW ON THE MATRIX SYSTEM**	106
4.1.	Morphology	106
4.1.1.	Possible Functions of the Acupuncture Point Organ	106
4.1.2.	The Question of Functional Relationships	107
4.1.2.1.	Changes in the Organ due to Stimulation of the Point	107
4.1.2.2.2.	Change in Physical Functions of the Point in Disease of the Associated Organ	108
4.1.3.	Phenomena of the Acupuncture Point That Can Be Palpated	109
4.1.4.	Thermal Phenomena	110
4.1.4.1.	Temperature Regulation Tests	110
4.1.5.	Electrophysiological Phenomena	110
4.1.5.1.	Conductance Investigations	110
4.1.5.2.	Experiments of Potential Differences	112
4.1.5.3.	Harmonization of the Rhythm	114
4.1.5.4.	Equipment and Methods	114
4.1.5.5.	Results	114
4.1.6.	Synopsis	115
5.	**DIAGNOSTIC METHODS**	116
5.1.	Palpation Reflex Signs of Illness	116
5.1.1.	Thermodiagnosis and Infrared Diagnosis	116
5.1.2.	Electrodiagnosis	117
5.1.2.1.	The Electro Skin Test	117
5.1.2.2.	Conductance Measurement	117
5.1.2.3.	Measurement of Difference in Potential	118
5.1.2.4.	Measurement of Capacity	118
5.1.2.5.	Measurement of the Body's Own Electromagnetic Signals	118
6.	**TOWARDS A REGULATORY THERAPY**	119
7.	**REFERENCES**	120

PART THREE — Felix Perger, MD, Editor

1.	THE THERAPEUTIC CONSEQUENCES OF MATRIX REGULATION RESEARCH	125
1.1.	The Puncture Phenomenon	125
1.1.1.	Bioelectric Events During Puncture Phenomenon	130
1.1.2.	Puncture Phenomenon and Oxygen Saturation of the Blood	132
1.1.3.	Iodometry and the Puncture Phenomenon	134
1.1.4.	Totality of Regulation During the Puncture Phenomenon	134
2.	TESTING THE INITIAL STATE AND AUTONOMIC ASYMMETRY	135
3.	EXTRANEURAL MECHANISM FOR CONTROLLING IMMUNE PROCESSES	151
4.	NEURAL THERAPY ACCORDING TO HUNEKE	160
5.	THERAPEUTIC CONSEQUENCES OF MATRIX REGULATION RESEARCH	172
5.1.	Selection of Specific Drugs and Avoidance of Delayed Effects	173
5.1.1.	Rehabilitation of Immune Capacity	175
5.1.2.	Conservative Therapy to Relieve Stress on the Immune Regulatory Cycles	176
5.2.	Elimination of Scar Disturbance Fields	179
5.2.1.	Release of Silent Chronic Inflammation (Foci)	180
5.3.	Follow-up Rehabilitation Treatment	181
5.4.	Success and Failure in Regulatory Therapy	182
6.	REFERENCES	185
	Index	199

Foreword to the 2007 English Edition

Extracellular matrix (ECM) regulation presents a holistic theory of the maintenance and recovery of health, as well as the development of illnesses. Practiced by the Viennese anatomist Alfred Pischinger (1899–1982) and his colleagues, ECM regulation represents a European system of investigation originally called the study of humors in the Hippocratic tradition. Galen (129–201 AD) more accurately understood this as humoral pathology, the study of blood composition. While Rudolf Virchow (1821–1902) replaced the humoral point of view with a theory of cellular pathology, Carl von Rokitansky (1804–1878) continued to adhere to the principle of humoral pathology and went on to refine Galen's principles with new insights into microcirculation. Von Rokitansky suggested that illness originated not in the cell itself, but in damage to microcirculation.

Viennese professor of internal medicine Hans Eppinger (1878–1848) further developed von Rokitansky's point of view using modern histology, experimental research, and clinical and pathological evidence. Eppinger showed that every cell requires a suitable environment called the ground substance—or the ECM—for successful microcirculation.

Pischinger continued this work by defining the connections between the ECM and both the hormonal and autonomic nervous systems. These scientific studies allowed Pischinger to draw connections between the brain and the ECM, as exhibited when the central nervous system registers situation-specific changes (somatopsychic information) and delivers a response to the periphery (psychosomatic information), impacting all of the connected organs, including ECM components. This communication network is now known as the system of extracellular matrix regulation.

This American edition brings together all the strands of Pischinger's ECM work so that it might be more widely known. It also contains my own new research and includes contemporary discoveries regarding the ECM: its construction, relationships to cybernetic non-linear systems, and phase transitions. The contributions of Dr. Bergsmann (Part Two) and Dr. Perger (Part Three) also represent new material.

The book follows in a great European medical tradition and will give physicians and therapists around the world an introduction to holistic medical thinking and treatment. It is a welcome endorsement on the part of North Atlantic Books to publish this work in the US. The volume represents an extraordinary achievement of Ingeborg Eibl, Shannon Kelly, and Lindy Hough. Susanne Seeger of Thieme Verlag, Stuttgart, Germany, repeatedly gave encouragement and facilitated communication. I thank all of them and hope this book receives the wide distribution it deserves.

PROFESSOR HARTMUT HEINE
Doctor of Natural Sciences
Spring 2007

Introduction to the English Edition

Regulation is a key topic for all areas of biomedicine and therapeutics. This book brings to the English-speaking world vital perspectives on regulatory biology. These perspectives have emerged from nearly forty years of research carried out by brilliant scientists and clinicians in Germany. While an English translation of this work has been done in the past (by Haug International), it has been out of print for some time. This new and updated version comes at a time when interest in the living connective tissue matrix and ground substance has greatly accelerated, owing to new research and clinical applications worldwide.

Stated simply, every function and every process in the living body involves the matrix in one way or another. The reason for this is that every cell in the body is nourished via the matrix, and all waste products of cellular metabolism likewise pass through the ground substance, which is the actual milieu. The matrix is also the *terrain* in which all immune responses and tissue repair processes take place. But there is much more to the story.

Of primary importance is the modern recognition that the matrix is not an inert filler substance or filter, but is instead a body-wide communication and support system, vital to all functions. For example, we now know that tensions in the matrix regulate metabolic processes in the cytoplasm and nuclei of cells (Ingber). A key discovery has been the identification of molecules called *integrins* that span the cell membrane, connecting the extracellular domain with the cytoskeleton and nuclear matrix (Bretscher). While the term *integrin* came after Pischinger's research, Pischinger recognized the bidirectional nature of information movement across the cell surface and its role in enabling groups of cells to react together.

Recent research has also demonstrated that the matrix components are actually semiconducting liquid crystals, materials known to have a variety of remarkable properties for the transmission, storage, and processing of information involved in regulations (Szent-Györgyi, Ho, Popp). Hence the book resolves long-standing confusions about the significance of holistic and complementary therapies. The collagenous matrix and ground substance of the human body form a totally pervasive system, a major organ, that reaches into every part and whose properties are absolutely vital to the operation of the whole. The matrix system has not been recognized by Western biomedicine as an actual organ because it is so intertwined with physiological regulations that it is challenging to study it or to even identify it as a separate system.

The systemic aspect of the matrix arises because of the continuously interconnected ground substance gel composed of proteoglycans and related molecules, with strong collagen fibers embedded within it. In no other book will you find such a detailed, dynamic, and clinically relevant picture of this system and the ways it maintains an electrostatic "tone," governing the availability of water, electrolytes, and electrons needed for proper functioning of all tissues. At the same time, part of this system called *the kinetic chain* conducts movements from muscles to bones, and acts in the reverse direction as the body's shock absorber. Some matrix components are key to flexibility, and others lubricate the joints.

A wide variety of complementary therapies have developed sophisticated methods for interacting with the ground regulation system. Both Western biomedicine and all of the so-called complementary therapies can find a valuable common denominator in the study of the ground matrix system. For it is in the matrix that the causes and cures of the so-called systemic and chronic illnesses can be found. This includes especially the inflammatory conditions that have become the focus of much modern medical research and that have been so difficult and costly to treat. Pischinger, Heine, and colleagues simply and brilliantly describe how disturbances of matrix regulation can simultaneously compromise homeostasis and efficiency, leading eventually to depletion and organ failure, here characterized as *regulatory rigidity, regulatory paralysis,* and *regulatory disintegration*. Pischinger's research provides details of the inflammation picture that are missing from most of the modern accounts. The principles are of more than theoretical interest; the practical clinical applications are described in Parts Two and Three.

Pischinger and his colleagues (Heine especially) recognized that the acupuncture meridians are the main channels of the matrix, and described the nature of acupuncture points in great detail. Their findings are congruent with very recent biophysical research on the subject (Jones, Langevin). This information will be valuable to all who use and research acupuncture.

This is more than a book about an important aspect of the living structure and function. It is a book about the philosophy of medicine; a book that explores the way scientific thought proceeds and emphasizes the importance of nonlinear perspectives. For one of the most important attributes of the living system, often forgotten, is that it is profoundly nonlinear and open to environmental influences, as recognized in the Nobel Prize research of Ilya Prigogine (1917–2003). Many treatments, and clinical trials based on them, rely on linear cause-effect models in which problems are viewed as isolated events and in which therapeutic substances are supposed to react with specific receptors on or within cells. There is no place in this overly simplistic scheme for vital system-wide changes to rapidly communicate through the entire matrix; nor is there an understanding that such system-wide communications can become disorganized to create a susceptibility to disease or to reduce energetic efficiency.

It is an intellectual challenge to explore nonlinear whole-system phenomena, but it is essential especially in the areas of prevention, environmental medicine, and the study of side effects, all of which document the disrupting influence of toxic substances that are increasingly part of our daily lives, and that can become bound within the matrix. Moreover, Pischinger and his colleagues recognized the optimistic aspects of this system: under the appropriate conditions the matrix can react quickly as a unit. Order and vitality can spread virtually instantly throughout the entire intermeshed system in an autocatalytic, or chain-reaction, manner. The proteoglycans in particular can react to every type of stimulus with a form of depolarization that can be rapidly propagated throughout the matrix system, as documented in a series of studies of the electrical changes occurring when the skin is punctured. Any stimulus exceeding a particular low-intensity threshold will trigger a reaction in the entire matrix.

Subtle stimulation at the appropriate points can elicit the release of long-held toxins and resolution of so-called silent inflammation. Such systemic changes have become part of our modern inquiries into homeostasis, cybernetics, synergetics, cooperative interactions, and psychoneuroimmunology. Responses of this kind have been routinely observed during the practice of homeopathy, an approach that Pischinger and Heine recognized and appreciated. Likewise, the effects of minimal stimulation of "singular points" (Shang) or active points located by sensitive palpation can create whole-system changes, as has been documented by acupuncturists for some thousands of years. These changes take place far too rapidly to be explained by slow-moving biochemical processes and the diffusion of hormones. In this book, the acupuncture point is referred to as the "window on the matrix system." The properties to look for when palpating for *active points* are described in detail. For students of acupuncture and its mechanisms, or for hands-on therapists, there is much to learn from this text. The German researchers have done a great deal of careful investigation of the electrical aspects of skin puncture and acupuncture.

The study of the matrix system and its rehabilitation is a field that begs for significant research and clinical application. For a difficult disease like cancer, for instance, rehabilitation of the ground-regulation system can improve outcomes. It is in the matrix that the solution to the cancer problem is most likely to be found. This book, particularly section 2.1, should be read by every oncologist and cancer researcher. The properties of the matrix demand a rethinking of the entire therapeutic process. Those who recommend antibiotics, antihistamines, pain medications, cortisol, chemotherapy, or any other therapeutic agent need to consult the information in this book for the best outcomes with minimal disturbance to the matrix and the myriad processes it influences. Pharmacological and surgical medicine should take note of this fact, and texts and manuals should reference this material. (Major interventions should not be undertaken, however, without first ensuring that the matrix and all related immune functions have been destressed. In other words, the success of any intervention depends on preparing in advance the proper conditions for a normal healing process.)

This new English edition—thanks to its editor, Professor Heine; its translator, Ms. Eibl; and its publisher, North Atlantic Books—is a valuable volume, and the most comprehensive picture of the living connective tissue matrix available in the English language today.

JAMES L. OSCHMAN, PHD
Author of *Energy Medicine: The Scientific Basis*

Foreword to the German Edition

The tenth edition of Arthur Pischinger's *The Extracellular Matrix and Ground Regulation* demonstrates the outstanding importance of this book for the theory and practice of biological medicine since 1974. This edition would not have been possible without the considerable commitment of Karl F. Haug Publishing House and its editor, Mr. C. v. Grumbkow, as well as many discussions with readers.

The system of matrix regulation can explain why every multimorbidity, every chronic illness, and every tumor problem begins with disruptions in the balanced state of the health of the individual (*Befindensstörungen*). Instead of realizing their immense importance to maintenance of health, these stressors are trivialized, and the old reliable medicines, labeled by insurers as uneconomical, are increasingly excluded from reimbursement. Disturbances of the balanced state of health (together with psychiatric, social, and economic problems) are now rarely viewed as illnesses, but are relegated to a position subordinate to illnesses, which can be objectively measured and put into scientific categories. The resulting "engineering character" of medicine comes at an enormous financial expense and only benefits a relatively low number of high-risk patients.

In contrast, the great numbers of people who have the beginning stages of systemic illnesses are ignored. This places a burden on the entire community. Similarly, so-called innovative medications and surgical procedures are becoming increasingly expensive. They are used not just where necessary, and their scope has even been extended to include use on patients who do not require them.

An exploration of the system of matrix regulation can clarify the meaning of the term "complementary medicine." This accords with the wishes of both Pischinger and other authors. Namely, that in the same way as individual parts need the whole, a purely causal explanation of life processes is incomplete without the wholeness provided by a complementary explanation regarding "meaning and purpose."

My good friend Otto Bergsmann died in July 2004. Along with him, Gisela Draczynski (d. 1998), Felix Perger (d. 1993), and I endeavored in the spirit of Pischinger to establish the system of matrix regulation in medicine on an international scale. Today, this effort has succeeded. Otto Bergsmann always emphasized that new findings should be linked with old ones. In Pischinger's memory we will continue to follow this path.

HARTMUT HEINE
Neuhausen, Summer 2004

Introduction

Experience teaches that it takes years or decades for new fundamental insights in medicine to become common knowledge among doctors. Additionally, any body of knowledge that reveals linear-causal relationships is more easily understood and categorized than those that reveal relationships which are multidimensional.

The problem of today's medicine lies in the fact that it sees linear-causal and multidimensional relationships as contradictory, instead of seeing that both can be used together to gain knowledge. The linear-causal and cause-and-effect types of thinking have made possible the extraordinary successes in both the fields of surgery and of acute illnesses. To a large extent, these successes have their basis in the cellular pathology of Virchow and in Newton's classical physics. Considered from this perspective, the random double-blind study is seen as the conclusive way to understand medical events.

However, it is obvious that chronic illnesses have become more prevalent in the last several decades. It should not be forgotten that the usual treatments for these diseases as well as their presumed pathogenesis are based on linear-causal thinking. This is the crux of the matter for chronic care medicine. Given this situation, developments in the field of cybernetics (Wiener 1963) as well as the thermodynamics of open systems have been quite useful in directing attention towards a solution.

These studies pointed out that biological systems, far from showing linearity, are highly interlinked and are subject to a biological flow equilibrium (von Bertalanffy 1952). As "open systems," they exchange energy and matter with their surroundings. In contrast to classical closed (mechanical Newtonian) systems, when these open systems are supplied with non-chaotic energy, it can immediately spread over the entire system. The important point here is the transmission and distribution of information. Taking cybernetic systems into account leads to more comprehensive knowledge, since it opens up a monocausal view into a multidimensional one.

In 1975, Alfred Pischinger, Professor of Histology and Embryology at the University of Vienna, presented his findings about the "System of Matrix Regulation" in the first edition of this book. In this work he described the "sounding board," upon which reactions in the human body play.

Pischinger was a student of such important teachers as H. Rabel, the Nobel Prize–winner [Otto] Loewi, Alberecht von Bethe, and Wilhelm von Möllendorf. With his first publications, including "A report on the isoelectric point of histological elements as the cause of their variations in capacity to accept staining," he became the founder of tissue histochemistry. From the beginning, he concerned himself with the communications which spread the connective tissue output over the entire organism.

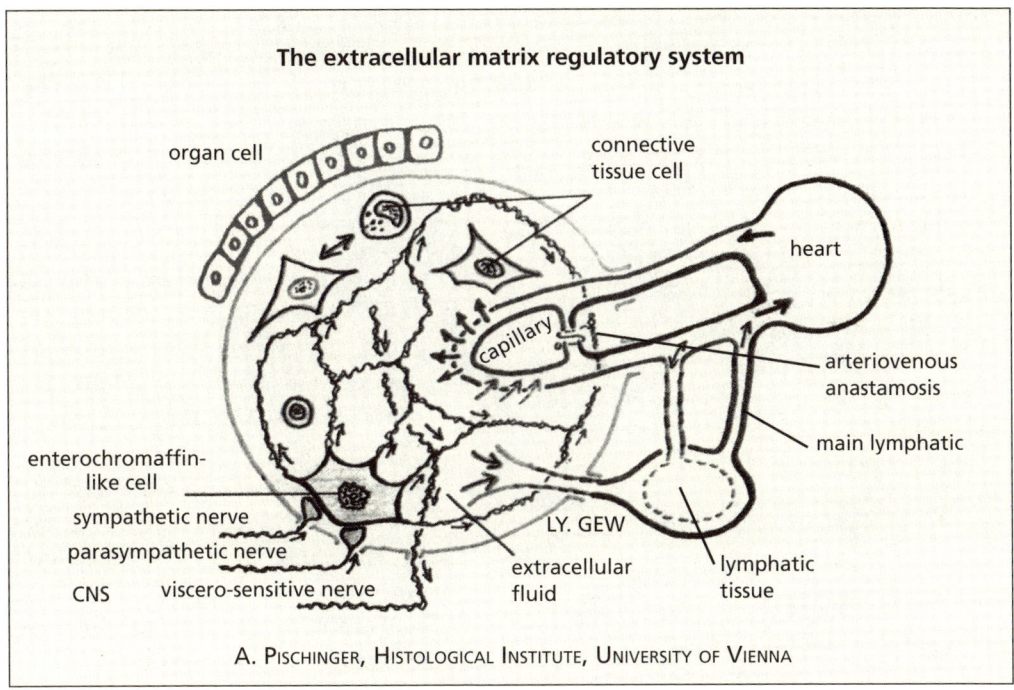

The extracellular matrix regulatory system.

Today, according to Pischinger, the system of matrix regulation is defined as a functional unit consisting of the capillary bed, the connective tissue cells, and the autonomic nerve endings. This triad has a common area of activity and information—the extracellular fluid. The lymphatics and lymphatic organs are connected with this matrix regulation system, which is the largest system that completely pervades the organism. The matrix regulation system manages the nutrition of the cells (inner circulation) and the removal of their waste products. Thus, it regulates the "cell milieu system" and, at the same time, is part of all inflammatory and immune processes. It is thus responsible for all basic vital functions.

The very existence of all organ cells depends on the intact functioning of the system that guarantees their environment. Organic diseases arise as a result of dysfunction of this interconnected, ubiquitous system. Over four decades, the influences on the matrix system of various noxious substances (silent chronic processes, heavy metal loads, the effects of stress, etc.) have been observed at the Vienna School using reaction measurement observations (Bergsmann, Kellner, Perger).

The dynamics of functional processes (normal function, types of dysfunction, and changes in the stimulation threshold) were recorded on thousands of patients by using various investigative methods.

In the process, a two-layer system of inflammation was revealed, which was classified as "nonspecific" and "specific."

The extracellular matrix regulation system can react locally or generally. Various stimuli set off similar types of reactions in the nonspecific part of the matrix regulation system. Hauss and Junge-Hülsing from the Münster School arrived at the same conclusions.

The Vienna School attributes the effective and lasting successes in understanding chronic diseases to two factors. First, for every single patient, attention was given to the different and individual stress factors that maintained the chronic disease state. Second, there was an understanding of the reaction processes of the matrix regulatory system.

Original experimental and treatment methods were developed to normalize the functions of the matrix system. In this way, chronic diseases that are generally considered difficult or impossible to treat (e.g., autonomic dysregulation, multiple sclerosis, ulcerative colitis, rheumatoid diseases, and, to some extent cancer processes), became amenable to treatment. Perger impressively proved the cost-saving effects of this approach. Focusing on the functioning processes in the matrix regulatory system opened up a genuine basis for prophylaxis.

Pischinger's system of matrix regulation has been shown to be both a teachable and usable concept. It encompasses all previous theories of medicine: humoral pathology, organ pathology, neuropathology, cell pathology, and permeability pathology. One of Pischinger's students, Gottfried Kellner, had a major role in the development and formulation of this concept. His works have also been published in book form by Karl F. Haug Publishers.

Three major schools, independently of one another, have recently been active in the field of matrix regulation.

1. Pischinger and his colleagues, who took over the tradition of the old humoral pathology of Galen and Paracelsus as well as more recently, von Rokitansky and von Eppinger. The clinicians Altmann, Bergsmann, Fleischhacker, Hopfer, Aiginger, Perger, Plohberger, von Riccabona and Stacher put the theory of matrix regulation into use in clinics and practices;
2. Hauss and Junge-Hülsing, University of Münster (universal mesenchymal reaction);
3. Heine, University of Witten/Herdecke.

With his fundamental research on the mesh structure of the matrix produced by connective tissue cells, Heine established the basis of information-exchange interactions that take place between cells, and this can be understood to apply to the entire organism.

After decades of experience with the system of matrix regulation that regulates the basis of life for all cells, it is becoming clearer that Pischinger was correct in speaking of a new foundation for a holistic biological theory of medicine.

At present, updating prior knowledge about the matrix regulatory system required a completely new edition of Alfred Pischinger's book. Professor Heine took on the redesign of the book in cooperation with Bergsmann and Perger, physicians who were personally and scientifically close to Pischinger. This new edition was only possible with the cooperation of the Pischinger family and with the particular interest of Karl F. Haug, Publishers, to whom we are deeply grateful.

GISELA DRACZYNSKI
Cologne, Autumn, 1988

PART ONE

1. Structure and Function of the Extracellular Matrix

The concept of a cell is, strictly speaking, only a morphological abstraction. Seen from a biological viewpoint, a cell can not be considered by itself without taking its environment into account.

With these words, Alfred Pischinger, as late as 1983, recognized the weaknesses of the paradigm of Virchow's then still-valid cell theory. In his work on cellular pathology, Virchow (1858) had limited the concept of illness to the disturbances in structure of the individual cells. His basic concept is that each of the approximately 50 billion cells in the human organism is an "elemental organism," existing primarily for itself alone, enclosed and bounded by the cell membrane, yet incorporated into a working organism, playing its part in the function of the whole.

The result of this linear, cause-and-effect thinking introduced into European natural science by Galileo (1564–1642), is that organisms are seen as analogs to technical machines, that is, complicated cellular functional units whose defects can correspondingly be repaired. In the end, what matters is finding the disease-causing molecule in the cell. (At present, it is believed that this has already been observed in spot mutations in individual amino acids.) The entire therapeutic system of academic medicine is affected by this linearity in medical thought, namely that a drug must attach itself to a suitable cell receptor, and a reaction can only occur if the reactants fit together like a key in a lock.

In attempting to cling to the simplistic viewpoint of cause-and-effect relationships, one has no choice but to separate the acute event from its intermeshed biological associations, call it a syndrome, and treat it as if it were an isolated event. Particularly in the case of chronic illnesses and tumors, it is evident that there is almost no noticeable difference between the short-term effect of certain treatments and their long-term effectiveness (Fülgraff 1985). Consequently, this causal, analytic linearity has influenced theoretical teaching in both clinical trials and medical treatment. Focusing on an illness according to its type replaces the individual phenomenon of being sick. Turning this disease into a model makes it accessible to instrumental measurements in a causal-analytical way. Reality is replaced by models, and the more complex the reality is, the more simplified the models become. "In this respect, medical experience is no longer cultivated, since models and not reality have become important." (Fülgraff 1985). A model offers neither the framework for determining individual biological outcomes nor for quality of life.

Additionally, given that the same symptoms can mask a variety of diseases, the very design of the randomized, double-blind clinical trial makes it only one way to obtain knowledge. It is actually inaccurate to insist that it is the only method, since case studies and narratives, based on the patient's detailed experiences, do precisely what is impossible for an "objective," controlled clinical trial to do—they primarily focus medical attention on the patient as an individual, and only secondarily on the disease. The reason that Virchow's cellular paradigm has been so successful in modern medicine, particularly in acute diseases and those caused by microorganisms, is because there are individual causes that can be objectified and can be eliminated or repaired immediately. However, at present this approach is seldom successful with the increasingly prevalent chronic diseases and tumors.

1.1. Functional Unit of the Cell and the Extracellular Space

Cells have a reciprocal relationship with their environment. Seawater is the primary regulatory system of the single cell. In multicellular organisms, the ion composition of the structured extracellular space corresponds to seawater. The milieu surrounding a single cell forms a structured ground substance—what we are calling the "extracellular matrix"—in multicellular organisms, and this matrix significantly determines the genetic expression of the cell (Hay 1983). From a macroscopic perspective, the organization of the extracellular matrix can be seen in the connective and supporting tissues as well as in the blood. But from a molecular biology viewpoint, the matrix is mainly concerned with sugar polymers—either in free form, or in a variety of forms with protein- and lipid-binding forms of intercellular substances, and the respective individual sugar membranes (glycocalyx) of the cell.

Bordeu (1767) already recognized that connective tissue has more than a mere supporting and filling function, and that it carries out nutritive and regenerative tasks in the service of specific organ functions, as well being a mediator of circulatory and nerve functions.

C. B. Reichert (1845) similarly understood that the connective tissue has a medium that is not only mechanically binding but organically vital (!). He also recognized that nerves and vessels do not come into direct contact with the functioning cells, but that the connective tissue substance is the *mediating* member, the bearer of of the nerve and nutrition flow, and that interactions pass through it everywhere in the body. Only the connective tissue has direct contact with all the parts of the body. However, at that time there was no test available that could show how the connective tissue could intervene in these interactions.

Until very recently, histological and biochemical techniques did not permit a more precise structural analysis of the extracellular matrix. At the same time, the more easily comprehended concept of "cells" exactly fit the model of the smallest building blocks of an organism, in the causal-analytical way of thinking popular at the beginning of the technical age. Only now, with knowledge of cybernetic relationships and open energy systems, are we beginning to reconsider this overemphasized, positivistic point of view. It is, of course, true that there were already early warnings about the dangers of a one-sided, model-type cell concept and the disregard of humoral factors (Rokitansky 1846, v. Rindfleisch 1869, Buttersack 1912):

It was bold and expedient of Virchow to follow local disturbances down to the level of the individual cell. With this, Virchow freed pathology from the vagueness of *one-sided humoral* and *neuristic* concepts. But he went one step too far when he specified the terrain. It is perfectly possible both to recognize the individuality of the cells and at the same time to remember the structures that limit their functional and nutritional autonomy. Certainly the cells of the parenchyma are excitable and active, but in this they are partly dependent on the nervous system. They feed themselves and grow, but in this they are partly dependent on the vascular system. The terrain where local disturbances take place consists of three parts: parenchyma, capillary beds, and nerve endings.

However, we cannot stop here. Even the way anatomical parts fit together is subject to certain general rules, and connective tissue plays a major role in this. In every location, *connective tissue* interposes itself between the bloodstream and the principal structural parts. In doing this, in its histological transformation, the connective tissue adapts to the individual needs of the site in a marvelously complete way. For our present purpose, we will only consider the position of mediator that the connective tissue takes between the parenchyma on one side and its blood and nerve supply on the other. It is known that physiology has little to say about such a *transmission function* of the connective tissue. But pathology is compelled to focus on a whole series of apparently random points, and awareness of these points is of great importance in understanding how tissues change during disease. As regards nutrition via the blood, pathologists place great emphasis on the *capillary membrane* as an *endothelial membrane that limits the connective tissue* (!).

However, do not imagine that supplying of the parenchymal cells with nutritive material is merely a general soaking and rinsing process, but rather think of it as a current of secretory fluids. This current leaves the blood near the cementing borders that unite the rhomboid endothelial cells into an interlocked membrane. Beyond that the current increases considerably through the mediation of a network of small fluid canals (canaliculi). This network is more or less sharply delineated from the adjacent matrix of connective tissue. At its intersections, the canalicular network contains nuclei with dependent protoplasm remnants, the so-called connective tissue corpuscles. In this way the nutritive fluid penetrates the spaces enclosing the functioning parenchymal cells, and becomes available to them. Afterwards, laden with the waste products of the parenchymal cells, the fluid is taken up into the openings of the lymphatic vessels, which are abundantly found in connective tissue. In summary—everything that comes out of the blood takes a somewhat complicated route through the connective tissue to the parenchymal cells and then into the lymph system. When I look into the future and muse about how often I and my readers will have to follow this path, how each part of understanding the local disease state is tied to the steps of this route, then I would like to call this the way to the knowledge of pathological histology.

Buttersack arrives at a similar point of view. He writes, for example, "that the layer that up to now has been called 'connective tissue' is not simply a better connecting medium between the organs. It does not simply move fluids in and out to mold a new parenchymal cell. It will become abundantly clear that if connective tissue is really a living part of the organism, then it has all the attributes of life and all basic functions."

Thus, [it is] not simply a "transit route." In view of the aspects outlined above, one must agree with Buttersack, that in most respects, the name "connective tissue" does seem to be an unsuitable one. However, it is not easy to choose a better name. "Perhaps the name 'basic tissue' comes closer to the mark. It has been long known that the tissue in question forms the basis of the entire body. Figuratively, it is also the ground in which all organs take root, and lastly, to a certain extent, both as ground and matrix, it has its own essential value." (Buttersack 1912)

Both Ricker's *Regulatory Pathology* (1924) and Eppinger's *Permeability Pathology* (1949) also recognize that the functional connections between the capillary bed and the cell are through the extracellular matrix, whose disruption is the starting point of diseases. Ricker showed that the many processes and presentations that are involved in healthy and pathological events of the physical body are nothing more than variations of one and the same basic physiological process—variations in metabolism and in energy exchange; variations of stage and of site ("law of stages"). With this, Ricker was able to confirm the experiential medicine passed on by Hippocrates, Galen, Paracelsus, and Hahnemann, according to which all disease processes are of the same type, and that the dose is the healing factor, not the specific substance. Hence, just as has been demonstrated by experience, a curative dose of one and the same remedy has the ability to cure many diseases, and it is possible for a curative dose of countless remedies to alleviate or cure one and the same illness.

All of the above authors presented observations and evidence to the effect that what is primarily affected in every type of disease process is the basic tissue—the extracellular matrix—and that it plays a significant role in healing processes.

This, then, is the viewpoint of the older literature: the sum total of the connective tissue, including the cells, the fibers and the interfiber masses, is accepted as a united complex that surrounds the specifically acting parenchymal cells and makes their maintenance and regeneration possible. In addition, this totality of connective tissue cannot be considered separately from the vessels, such as capillaries, or from blood and blood formation centers. Thus, some of the new descriptions of changes in connective tissue such as desmoses, collagenoses or geloses are misleading insofar as they always consider only *one* of the components of connective tissues—either the fibers or the interfiber fluid. One should not fail to recognize that the cell is the central point in the connective tissue as well. The cell and its reactions, which have been long known to pathologists, are the visible expression of the autonomic activity and reactivity of the connective tissue.

In terms of the impact of this era of research, the result was continued interest in connective tissue, but these works were more concerned with its physiological performance and its biological tasks.

Before Virchow's cell theory became dominant, the individuality of each disease was considered to be an alteration in bodily fluids. This concept dates back to ancient times. Alkmaion, Galen, Hippocrates, and their students differentiated among four bodily fluids (humors): blood, mucus, yellow and black bile. The proper mixture (eucrasia) of these humors was the foundation of health, and a disturbed mixture (dyscrasia) was the basis of disease. The fluids were seen as the support of the physical constitution. One of Virchow's contemporaries, the Viennese pathologist Rokitansky, expanded the humoral theory into a major theory. Because the humoral pathology of that time was unable to demonstrate the same concrete evidence as Virchow's cellular pathology, it was displaced in the ensuing years. Beginning in 1945 in Austria, Pischinger and his coworkers were the first to give a rational medical/natural sciences methodology to the humoral theory as the basic system of matrix regulation. They took the concept of "cell," isolated by Virchow, out of abstraction, and placed it into a triad: capillary, extracellular matrix, and cell. This triad is the smallest *functional* common denominator of life in a vertebrate organism.

At approximately the same time, the emergence of cybernetics (Wiener 1963) and the development of the theory of "thermodynamically open energy systems" demonstrated that biological systems do not show linearity. They are highly interlinked and are subject to a balance of biological flow (von Bertalanffy 1952). This means that biological systems are energetically open and are thus in a position to exchange energy and matter with their surroundings. However, the arrangements that appear are not stable. They swing away from a thermal balance, which generally does not permit a return to the initial state. (Despite this, as in the case of genetic material, a stability can be reached that otherwise only belongs to minerals). In contrast to the classical closed, so-called Newtonian systems, open systems show that when suitable energy is supplied (non-chaotic energy, e.g., nutritional substances), it can instantly spread throughout the entire system. In an autocatalytic manner (acting as its own catalyst), this leads to the appearance of new structures, and these can also develop further into structures on a higher order.

Once cybernetic associations are taken into account, one is forced to abandon the firm ground of monocausal thinking. For the most part in biological systems, there is no obvious causal relationship between the guiding input on the one hand, and the end results (e.g., underestimated side-effects of drugs) on the other. "Anyone, however, who tries to apply one-dimensional causal chains to intermeshed systems can no longer claim to be scientific." (Thomas 1984)

Thus the difficulties in finding linear cause-effect relationships in the organism lie in the fact that highly intermeshed, energetic open systems are involved. The most appropriate form of energy to give structure and organization to a biological system, and to maintain that structure and organization, is the input and processing of information. The major significance of information as a non-chaotic energy form is that it is not tied to any particular energy carrier (e.g., sound waves in the air become information which is transmitted by the auditory ossicles in the middle ear, then to the sensory cells in the cochlea, from there to the eighth cranial nerve, and finally the information is transmitted to the appropriated neuron areas in the brain). So, in the living system, information is the most suitable energy carrier for triggering local as well as far-ranging interactions. This accords with the basic objective of an organism, that of keeping itself intact. It is certainly necessary to diagnose individual conditions in a causal-analytical way, but since chronic diseases are on the increase, there must be a search for a higher-level organizing principle which serves as the basis for the struggle to maintain the organism.

This principle is provided by the extracellular matrix and its regulatory mechanism (Pischinger 1983). Accordingly, the extracellular matrix permeates the extracellular spaces of the entire organism, reaches every cell, and always reacts as a unit. In epithelial cell groups or in the brain where extracellular space is reduced to the minimum, the extracellular matrix forms the intercellular substance. Biochemically, the extracellular matrix forms a meshwork of high-polymer sugar-protein complexes, with the proteoglycans predominating (Figures 3), followed by structural glycoproteins (e.g., collagen, elastin, fibronectin, laminin). Proteoglycans (PGs), glycosaminoglycans (GAGs) and structural glycoproteins form a molecular sieve through which the entire metabolism of the capillaries going to and coming from the cell must penetrate (transit route). Molecules over a certain size and/or charge can be excluded. The pore size of the filter is determined by the existing concentration of proteoglycans in the tissue compartment concerned, and by the PG molecular weight as well as by the electrolytes and resulting pH value. Here the negative charge of the proteoglycans has a crucial functional significance, and makes it capable of binding water and exchanging monovalent cations for bivalent cations.

Thus proteoglycans guarantee ionic, osmotic, and tonic homeostasis in the extracellular matrix (Hauss et al. 1968). The basic electrostatic tone established by this reacts to every change in the extracellular matrix with fluctuations in potential. The information encoded in this way can be passed on to the glycocalyx of the cell membrane as fluctuations in potential and, if they are strong enough (information selection!), lead to a cell reaction through depolarization of the cell membrane (e.g., muscle and nerve cells)—or, as with all other types of cells—through activation of secondary messengers (cyclic adenosine monophosphate, inositol triphosphate, etc.) on the membrane which transmit information coded in the extracellular matrix to

Diagram of matrix regulation

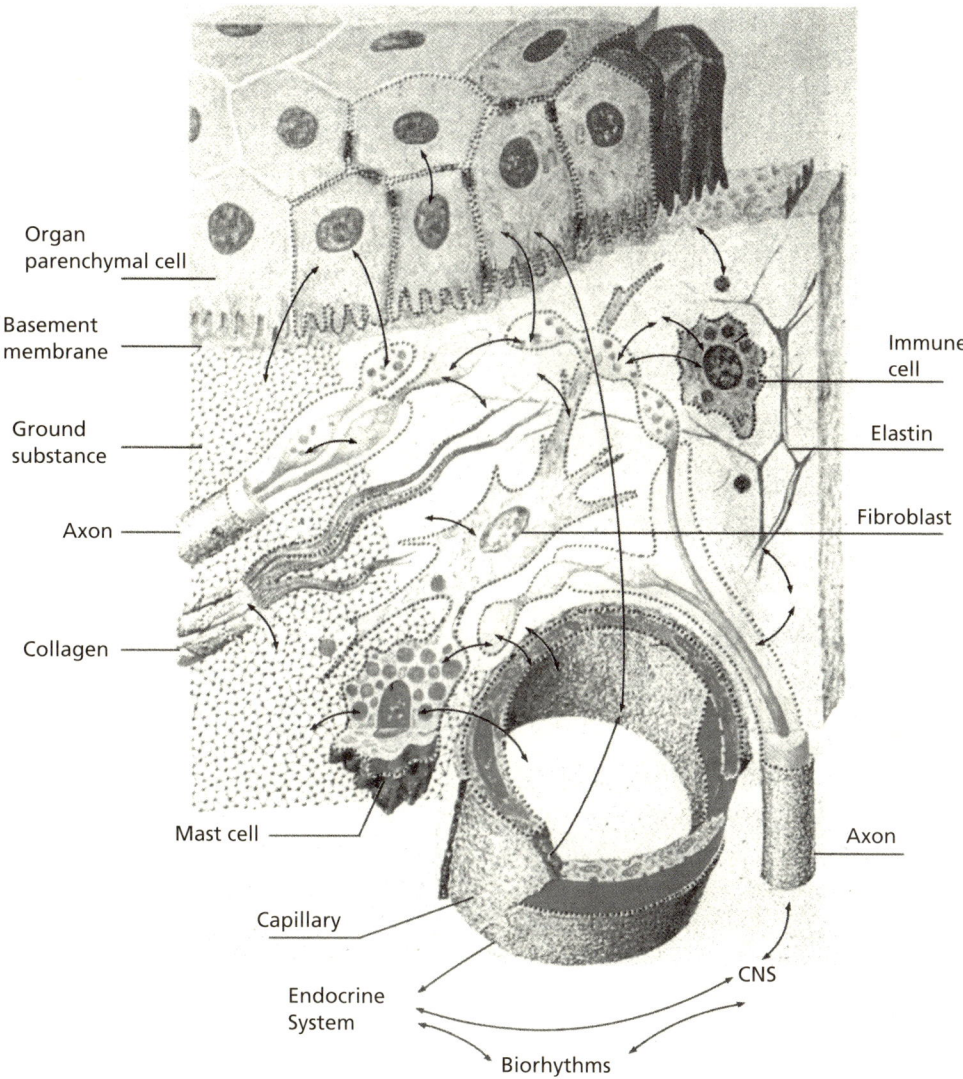

Figure 1: Diagram of matrix regulation. Reciprocal relationships (arrows) between capillary bed (capillaries and lymph vessels), extracellular matrix, terminal autonomic axons, connective tissue cells (mast cells, immune cells, fibroblasts, etc.), and organ parenchyma cells. Epithelial and endothelial cell groups are supported by a basement membrane mediated by the extracellular matrix.

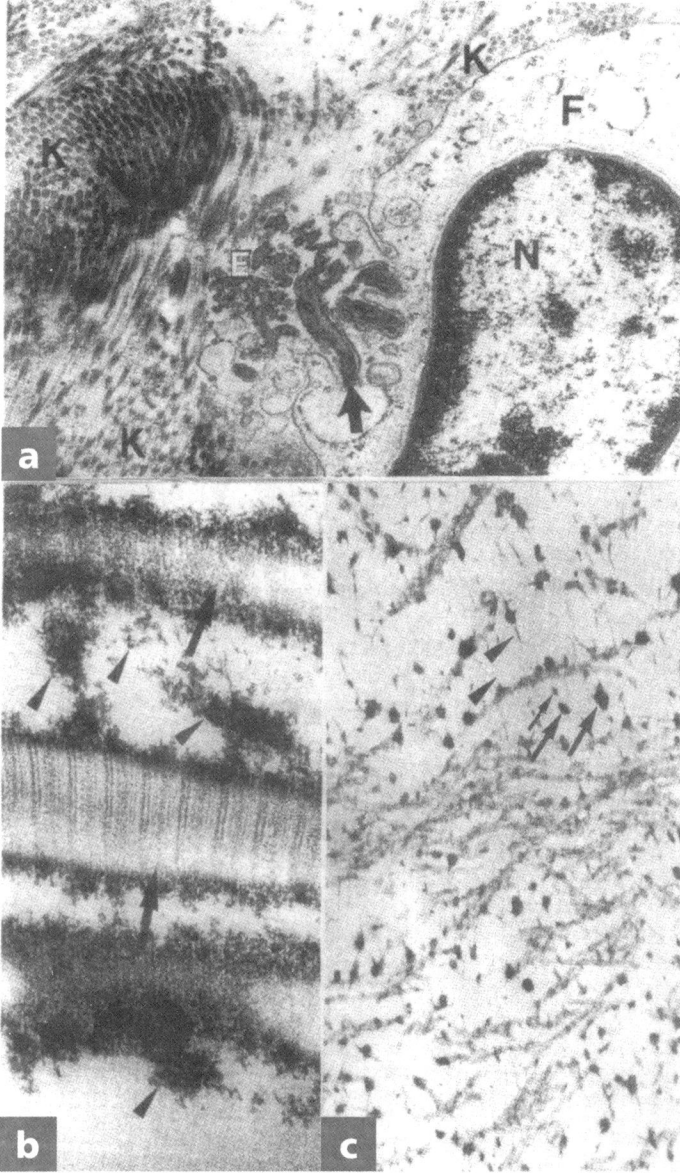

Figure 2: Ultrastructure of the extracellular matrix and matrix synthesis (compare with Fig. 1)

a) Human skin. Section of a matrix-synthesizing fibroblast (F). The arrow points to the extrusion of an elastic fibril. Collagen (K) and elastic fibrils (E) in the ground substance and and in the cell membrane. The proteoglycan mesh is recognizable as a delicate veil between the collagen fibrils. (N) = nucleus of the fibrocyte. X 20,000.

b) Human skin. View of the striations of collagen fibrils (arrows) and of the proteoglycans and glycosaminoglycans (PG/GAGs) (arrowheads) which are bound to the fibrils. The preparation was pretreated with ruthenium red (after Luft) to enhance the sugar and afterwards fixed in osmium tetroxide. Further treatment of the specimen was conventional. X 75,000.

c) Cartilage ground substance of a chick on day 21 of development. The mesh of the ground substance can be clearly recognized. The electron-dense 'buttons' are the proteoglycans (arrows), which collapsed because of the fixing procedure. The side arms are hyaluronic acid (arrowheads). The weakly electron-dense fibrillated structures are the structural glycoprotein chondronectin. X 56,000.

the cytoplasmic enzymes. These arrive in the cell nucleus and finally activate the appropriate place in the genetic material. This is followed by transcription of the corresponding DNA segment (gene) into the various types of RNA. After this transfer into the cytoplasm, different types of RNA in the channels of the endoplasmic reticulum start translating the information into products specific to the cell (Summary, Heine and Schaeg 1979).

The mechanical cohesion of tissues is largely determined by the mesh-type macromolecular superstructure of PG/GAGs (Balasz and Gibbs, 1970, Buddecke 1971). Through this, for example, the terminal axons of autonomic nerve fibers come under a very specific mechanical and electrical tension, and can react by releasing neurotransmitters and neuropeptides. PG/GAGs form a shock-absorbing system that acts as a lubricant (synovial fluid) that changes into a visco-elastic substance with severe and repeated mechanical demands. Because this is highly and elastically malleable, it consumes energy. Consequently, biochemically-coupled rheological changes are also part of the encoding of information in the extracellular matrix (Heine and Schaeg 1979, Heine 1997).

Every cell surface has a glycoprotein and lipid film (dotted line) which is bound to the matrix. There are also histocompatability complexes (MHC) on this film. The matrix is connected to the endocrine system through the capillary bed, and to the central nervous system via the axons. The fibroblast is the metabolic active center (Heine 1979).

1.1.1. Metabolism in the Matrix

The microstructural arrangement of the ground substance components can be described as a meshwork of similar polygons (Fig. 1). Their diameter varies between about 5 and 80 mm. Even if artifacts are produced while processing this meshwork, an inherent structural principle of the ground substance called the "matrisome" becomes clear. Naming it matrisome simplifies this structural principle and makes it comprehensible without ignoring functional details (Grimaud and Lortat-Jacob 1994). The matrisome is assembled from four macromolecules (Fig. 4): PG/GAGs, structural glycoproteins, reticular glycoproteins, and variable transitionally-bonded glycoproteins (cytokines, growth hormones, hormones, neurological substances, metabolites, and catabolites). These variable glycoproteins influence the function of the extracellular matrix.

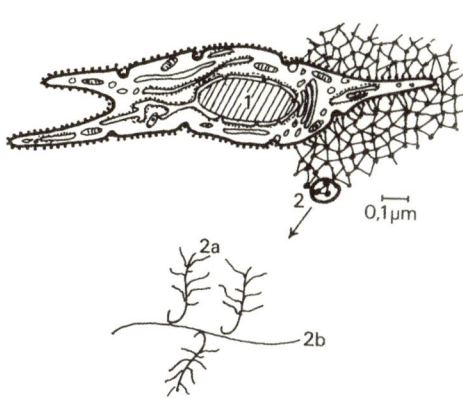

Figure 3: Section from Fig. 1. Matrix-synthesizing fibroblast (1). The proteoglycan meshwork pattern (2) is enlarged (arrow). Proteoglycans (2a) are bound to hyaluronic acid (2b) in the extracellular matrix.

The similarity of the matrisomes to each other shows that the ground substance is structured as determinate chaos, in other words, that labile, determinate structures are far from being in a thermodynamic equilibrium. Similar corresponding polygons originate from straight lines that intersect each other in all directions. Reconstructing the ground substance three-dimensionally by superimposing computer-simulated grids of lines randomly on the ground substance makes a specific layering appear that is not simply an impenetrable mat. This layering is organized in the manner of an assembly-disassembly process which gives rise to twisted hyperbolic forms (Fig. 5) whose structure is consistent with the principle of energetically minimal surfaces, a comprehensive summary of which can be found in Heine 1997. Here "minimal" does not refer to the size of the surface, but rather to the potential energy of the surface. Energetically-minimal surfaces are shaped like a negative gauss curve, such as the shape of a saddle (von Schnering 1991).

In vivo there is no evidence of entire energetically minimal surfaces since they are subject to constant deformations (e.g., through pH value changes, metabolic products, and so on). Minimal surfaces have the remarkable quality that, as a rule, they react to small changes in one place with very large changes at a great distance (Karcher and Polthier 1990).

The principle of energetically-minimal surfaces as space-dividing structural elements is widely distributed in nature. It is found repeatedly in the construction of blood vessels, nerves, tendons, bones,

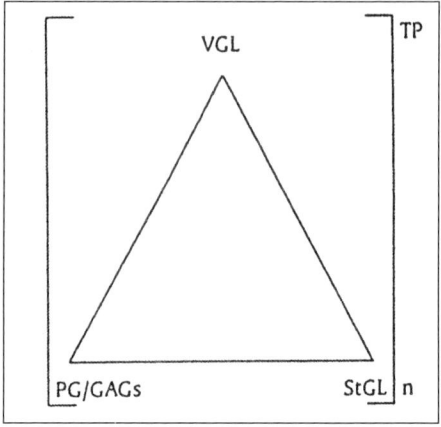

Figure 4: The matrisome is comprised of four macromolecules: proteoglycan/glycosaminoglycan (PG/GAGs); structural glycoproteins (St. GL); intermeshed (interactive) glycoproteins (VGL); temporarily bonded glycoproteins (cytokines, growth hormones, hormones, neurological substances, metabolites, catabolites, etc.). These temporarily bonded glycoproteins influence the function of the ground substance. The brackets (n) indicate the repetitive, nearly identical form of the matrisomes in the extracellular matrix (altered, after Grimaud and Lortat-Jacob 1994).

joint surfaces, cell membranes, DNA, enzymes, among others. Nonbonding interactions induced on the basis of the minimal surfaces, can influence the energetic requirements of all chemical reactions, including transmembrane transport, antigen-antibody interactions, protein synthesis, oxidation reactions, actin-myosin interactions, sol and gel states of polysaccharides (Andersson et al. 1988, Karcher and Polthier, 1990, synopsis by Heine 1997). The shifts in energy that course along the hyperbolas through nonbonding interactions open up new potential interpretations for many therapeutic methods of biologic medicine such as homeopathy, electroacupuncture, bioresonance experiences, and bioenergetic functional diagnostics. However, because under energetically minimum requirements the energy of only one photon which has no mass can have considerable effects, it is the energy that plays a role, and not the number of molecules (e.g., Loschmidt's number) or the law of mass action (synopsis by Heine 1997).

The potential ring terminals on the sugar components of the PG/GAGs appear to have significant control over the dynamic shape of tunnel-type hyperbolas, which are in the range of nanometers (c. 5–80 nm). For normal and/or disturbed matrix regulation to take place, the hyperbolas must be capable of building enclosure complexes that are called "guest-host" complexes. On the inside of the tunnel, hydrophobic (liphophilic) substances can be bonded to the hydrophilic substances of the outer wall (Fig. 5). This is the premise behind simultaneous transport of hydrophobic and hydrophilic substances through the ground substance (Heine 1997).

The extracellular matrix is connected into the endocrine gland system through the capillaries, and into the central nervous system by the peripheral autonomic nerve fibers which end blindly in the extracellular matrix. Since both systems are connected to each other in the brainstem, the higher regulatory centers can be influenced through the extracellular matrix. Capillaries, nerves, and connective tissue cells that regulate the ground substance can reciprocally (paracrine) be influenced through wandering connective tissue cells—macrophages, leukocytes, and mast cells. In addition, these same capillaries, nerves, and connective tissue cells can be influenced through "information" from released cell products (prostaglandins, cytokines, proteases, protease inhibitors, etc.) via self-feedback (autocrine). This creates a vastly complex, interlinked humoral system whose scientific predecessor exists in the classical humoral system. The advantage of this type of interlinked system lies in a considerable increase in adaptability and capabilities, as well as in the possibility of the appearance of completely new qualities that could not possibly have been achieved simply by adding the single properties of the components. In this

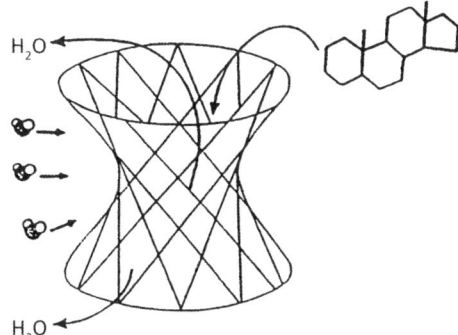

Figure 5. Tunnel structure of PG/GAGs of the ground substance. Bonding and reactions between the hydrophobic inner surface and the hydrophobic organic substance (guest-host complex) lead to the expulsion of water, which binds to the outer surface of the tunnel.

way, relationships between the psyche and the immune system (psychoneuroimmunology) become understandable. (Ader 1981, details and summary by Heine 1997.)

Despite the high degree of specialization in the body, such as the immune system, which creates a conditional susceptibility to faults, highly interlinked biological systems are redundant, and this is useful in evolutionary terms. This means that "the system compensates for the failure of individual components or subsystems because, if necessary, at times other components or subsystems can repair the defective parts completely or partially, some or all of the time" (Thomas 1986). Consequently, the basis of the ultimate goal of an organism, to survive, lies in the regulation of homeostasis, Therefore, in biology as well as in medicine, causality and final purpose are not mutually exclusive, but are interdependent.

In terms of biological evolution, the extracellular matrix is older than the nervous and hormonal systems. Accordingly, the formation and destruction of the extracellular matrix as a compensation mechanism is guided by a very primitive cell system: the fibrocyte-macrophage system. As the situation warrants, fibrocytes are capable of reacting within seconds, and produce a synthesis of proteoglycans and structural glycoproteins appropriate both in amount and quality. At the same time macrophages can normally break down the ground substance through phagocytosis. However, since the fibrocyte cannot distinguish between "good and bad," if such changes become chronic, an increasingly un-physiological ground substance is formed, which can lead to the development of a range of dysfunctional conditions from chronic illness all the way to tumors, through influence of the matrix on all cellular elements (Heine 1987b). The sugar polymers of the extracellular matrix are suitable for information transmission and information storage because they have strong capacities for binding water and ion exchange. Except for the biopolymer DNA, which perpetuates the genetic code, the sugar polymers in the ground substance are not involved in storing information to potentially transmit it through transcription and translation, but rather, they are involved in quick and orderly information transmission and distribution in order to actually regulate homeostasis.

In my opinion, the structural combinations of water and sugar bipolymers represent the oldest information systems and immune systems of aerobic unicellular and multicellular organisms (in unicellular organisms, bacteria, and viruses, the sugar polymers are bound to the cell membrane as the outer hull). In addition, at the level of homeostasis these polymers are able to help regulate the latent inflammatory preparedness of the connective tissue as the redox system, by taking and giving up electrons. (Levine and Kidd 1985). Because of these redox qualities, every change in the electrical tone, due to changing states of the extracellular matrix, can be coded as information and transmitted and processed in extensive interactions throughout the organism. At the same time, excess extracellular electrons and protons in the form of oxygen and hydroxyl radicals, which appear in every enzyme-guided transfer, can be intercepted by the water and sugar polymers. The resulting heat is in turn necessary to stimulate biological processes. The regulatory capability of the extracellular matrix thus has major significance in disease processes. In all acute and chronic diseases, including tumors, regulatory disturbances and changes in the ultrastructure of the matrix can be demonstrated (Pischinger 1983, Perger 1990, Heine 1987).

In the course of evolution, sugar polymers of the extracellular matrix become bound to a protein core, hence the term "proteoglycans" (an exception is hyaluronic acid, Fig. 3, 6, 7); or these polymers became bound to the outer surface of the cell membrane by membrane proteins and membrane lipids (glycoproteins and glycolipids of the sugar surface film—the glycocalyx—of the cell). All structural proteins (collagen, elastin, fibronectin, etc.) also became glycosylated.

Sugars in the form of nucleotides as a part of coenzymes are involved in most of the enzymatic reactions in the extracellular matrix and within the cells. Nucleotides are comprised of a nitrogenous base, a monosaccharide (almost always a ribose), and a phosphate group. Precisely because coenzymes mediate between various enzymes, they have a special significance in metabolism as a connecting link which enables metabolism to occur in the first place.

Nucleotides, as the term indicates, were initially discovered as a building blocks of the nucleic acids (DNA, RNA). In addition, certain secondary messengers such as cAMP, cGMP, and inositol phosphate that mediate extra-intracellular information transfer contain mononucleotides. An ancient pre-cellular evolutionary event seems to be reflected in this persistent "sugar principle of the living." The water-sugar-biopolymers have always remained evolutively modern.

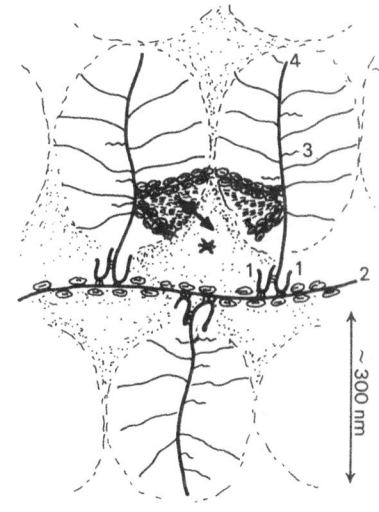

Figure 6. Detail from Fig. 3. Link proteins (1) bind the proteoglycan molecules to hyaluronic acid (2), which is stretched out because of its negative charge. The same happens to polysaccharide chains (3) which extend out from the protein core (backbone) (4). The broken lines depict the "domain" of a proteoglycan molecule. The double arrow points to fluid crystalline water and the ion exchange capacity (*) between the polysaccharide chains.

The Janus-like dual nature of this evolutionary step immediately became clear once oxygen appeared as the basis of life. On one hand, it is necessary for more highly organized life forms to obtain

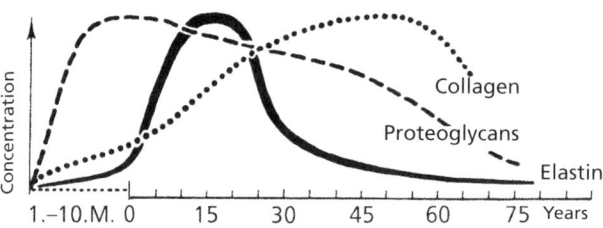

Figure 7. Progression over time. Synthesis of the most important macromolecules of the extracellular matrix (proteoglycans, collagen, and elastin) (after B. Robert and L. Robert 1973).

energy from oxygen through the metabolism of ATP along the mitochondrial respiratory pathway, and on the other, the resulting inflammation-promoting radicals must be rendered harmless. The energy released in anti-oxidative enzymatic processes can be absorbed by the water-sugar polymers of the extracellular matrix. This not only cools down the organism's "reactor," but at the same time makes available the energy needed for maintaining homeostasis.

Similar events occur with high-energy electrons which originate, to a significant extent, from the oxidative breakdown of carbon-hydrogen bonds, such as during the cleaving of glucose (Levine and Kidd 1985). In the course of evolution, important intracellular and extracellular antioxidant systems developed, such as intracellular superoxide dismutase, catalase, glutathione peroxidase, and, in the extracellular space, ascorbic acid, vitamins A and E, among others. The electron and proton displacements that appear in enzymatic oxygen metabolism lead mainly to the formation of multiple radicals. Their energy is stored in the physiological redox potential of the organism via the ground substance. If the enzymatic steps responsible for electron and proton transfer are disturbed, which can primarily begin focally (for instance, through inadequate blood supply), the result is an accumulation of free radicals.

If this continues over time, the resulting non-physiological alteration of the redox potential of the extracellular matrix leads to the risk of developing chronic inflammatory diseases, and even to developing tumors (Pischinger 1983, Perger 1990).

1.1.2. Chemical Sensitivity and Environmental Medicine

Toxins come from the environment and are also found in the increasingly prevalent man-made organic compounds. These latter compounds are found in such places as foodstuffs, where there are approximately 1,000 man-made molecules in use; in stimulants such as coffee, tea, tobacco; in drinking water, and in outgassing from paints, lacquers, and artificial floor and furniture coverings. These toxins show up in the body, where they are bound by the extracellular matrix and thus they can become pathological information for the cells that are downstream.

This pathological information can also manifest as the symptoms of chemical sensitivity (chemical sensitivity summary found in Rea 1995).

Since psychic stress can also lead to an increase in free radicals in the extracellular matrix, burdensome spiritual-psychic situations add to the stress load of the body.

This eventually leads to a maladaptation phase of matrix regulation. This situation commonly goes unrecognized, because it is mistakenly assumed that, when someone has lived with an environmental load and has remained outwardly healthy, a physiological adaptation—so-called "increased resistance"—has taken place. If a sensitivity is created by an overabundance of a man-made substance that an individual previously tolerated, this sensitivity can quickly spread to other substances (spreading effect). A heightened sensitivity to even minimal doses of the agent can then develop (Rea 1995). Removing the substance can be followed by a rapid recovery. If the patient is then challenged with the offending substance—under controlled circumstances—the symptoms reemerge. A long-lasting therapy is possible with an appropriate preparation of the *Noxe* (noxious substance) in such forms as homeopathic preparation, nosodes, or vaccines. (summary in Rea 1995).

Often, in a person around the age of fifty, an established maladaptation becomes apparent as multimorbidity (several concurrent medical conditions) or a chronic illness. Screening of the matrix regulation, e.g., testing for electrolytes in whole blood, acidosis, Methode Vincent, Decoder Dermography, EAV (Electro Acupuncture Voll), and blood cell values can give indications about regulation deficits. If the "matrix regulation cup" runs over, it no longer matters whether an emotional, chemical, or another stress caused the overflow (Ausschlag); the event uniformly discharges into degenerative symptoms (overview in Heine 1997).

1.2. The Extracellular Matrix as a Protein Regulator ("Slag Phenomenon")

Too little consideration has been given to the fact that proteoglycans (PGs) are able to store all four nutritional groups as follows: carbohydrates as glucose and galactose, protein as NH groups, fat as carbohydrates with oxygen esters ('fatty acids'), and water in the domain of PGs (Wendt and Warning 1986). Of all of these nutritional substances, water is the most important. When the proportion of water decreases, the brush-shaped PG forms fold together, and the transit routes in the extracellular matrix no longer function properly (Figs. 1, 3, 6).

However, these details do weaken a fundamental principle of currently accepted nutritional theory, namely, that the only place that humans can store protein is in their fat cells, where the excess calories are stored as triglycerides (Rapoport 1969). Upon closer examination, in addition to the fat cells, one finds there is an increased proportion of collagen in the connective tissue of obese people (Wendt 1984). Since collagen fibrils need polysaccharides for side-to-side polymerization (overview, Hay 1983), these are also increased. Protein can thus be stored in the form of collagen, proteoglycans, and glycosaminoglycans. Thus the entire ground substance of the organism has the capacity for storing protein, with certain organs having preference.

Excess carbohydrates are stored in muscle and liver cells in the form of glycogen, but they also lead to an increased formation of PGs in subcutaneous and interstitial connective tissue. Protein deposits on the basement membranes of capillaries and in the extracellular matrix have a higher sugar content in diabetics than in non-diabetics (Wendt and Warning 1986). This finding confirms that carbohydrates can be stored as PGs. In addition, sugar can also be bound by non-enzymatic glycosylation of proteins (for example, HbA1c), collagen, elastin, proteoglycans, albumin, myelin, cell membranes, etc. This reaction obviously has great significance in the aging process, the genesis of arteriosclerosis and in tissue changes in diabetics (Wendt and Warning 1986, Cerami et al. 1987).

While in stored protein the ratio of collagen to polysaccharide is 95:5; in pathological glycoproteins such as amyloid, the ratio is 42:58 (Wendt 1984). Stored protein and amyloid in changing quantities and combinations can bond to many other molecules known as "slag formation," such as immunoglobulins, lipoproteins, fibrinogen-complement, albumin, amino acids, glycoproteins, defective proteins, foreign antigen proteins, uric acid, cholesterol, environmental noxious substances, and carboxyhemoglobin. Hence, the alimentary tract is not necessarily the only place in which protein storage diseases can originate. The heteroproteins that arrive in the bloodstream are stored partly in the basement membranes and partly in the extracellular matrix. If the stored substances exceed the various capacities for their breakdown, they are increasingly shifted into the transit routes. This causes inflammatory micro- and macro-angiopathies, for example, in smokers, through carboxyhemoglobin deposits in their vessel walls, which can lead to intermittent claudication in the legs due to endarteritis obliterans, or to death from a myocardial infarction due to coronary arteritis (Wendt and Warning 1986).

At any rate, deposits of metabolic waste ("slag") which can be reduced, for example, with protein fasting. However, there are major individual differences in regenerative capacity, e.g., through the potential for hydrolytic and proteolytic extracellular breakdown by lysosomes released from physiologically lytic, neutrophylic granulocytes, or through intracellular digestion by phagocytic macrophages.

Each individual case must be examined to determine the extent to which genetic and extracellular influences (disposition and exposition) act together. This is demonstrated in Alzheimer's disease, which is associated with amyloid deposits in the brain. A precursor protein, coded through a gene on chromosome 21, has been shown to be involved in the formation of amyloid. Correspondingly, in Down's syndrome (trisomy 21) there are amyloid deposits in the brain in mid-life. Another example is a peculiar parallel between protein storage disease and hereditary familial hypercholesterolinemia. In both these conditions the numbers of LDL receptors are reduced due to genetics; here the mutations are on the long arm of chromosome 11 (Hobbs 1987).

Because of this mutation, the liver cells only have a few receptors which can attach LDL molecules and take them inside the cells. The LDL level in the blood is high because the LDL receptors are scarce, and the danger of arteriosclerotic changes is high as well. Getting the LDL molecules inside the liver cells requires cholesterol as a transport medium, and this cholesterol is produced in the liver according to information from the receptors. So, despite the high LDL level, the liver cells attempt to balance the situation by synthesizing more cholesterol.

In patients with protein storage diseases, the transit route between liver sinusoid and liver cell membrane (space of Disse) becomes increasingly blocked up, so that the LDL feedback to the liver cells likewise drops too much, and then the same effect occurs as in the case of genetically-caused lack of receptors, namely, the synthesis of LDL cholesterol increases (Wendt and Warning 1986). In contrast to this, the recessively inherited mucopolysacchridoses with their multifaceted connective tissue weaknesses make up a good straightforward group of conditions (Lenz 1983).

1.3. The Influence of Matrix Vesicles on Ground Regulation

A little-noticed yet important regulatory principle of the extracellular matrix in normal physiological as well as in pathological relationships is the pinching-off (and shedding) of vesicles from the connective tissue cells and immune cells into the extracellular matrix (Heine 1987a, b; Fig. 8, 9). There they disintegrate because of the release of a large number of biologically active substances (e.g., proteolytic and hydrolytic enzymes), as well as a large number of cytokines (e.g., prostaglandins and leukotrines) which also come from the disintegration of the vesicle membranes.

This disintegration of matrix vesicles not only influences the pH value of the tissue, thereby affecting the regulation of homeostasis in a nonspecific manner; it also nonspecifically influences cell and nerve functioning in the way that single-celled glands do, in autocrine (self-influencing) and paracrine (influencing the immediate environment) ways. (Heine 1987a, 1988c) The phenomenon of normal physiological lysis (disintegration) of white blood cells, described by Pischinger (1957) and Kellner (1976) also partly regulates the matrix. The enormous quantity of cytokines, lymphokines, and tissue hormones released in this process is eminently able to activate the highly intermeshed extracellular matrix in many locations and in many ways simultaneously. Of necessity, this activation must take place in a nonspecific manner, since this is the only unerring way to stimulate the self-healing powers of the organism.

Thus, the wheel that turns all natural medical treatments and regulatory medical procedures is the normal physiological ability of the white blood cells to disintegrate (Heine 1988c). Pischinger rescued this phenomenon from oblivion when he sought an explanation for the phenomenon that within 24 hours more lymphocytes enter the bloodstream than the total number that are present in the blood (at any one time). In humans, the influx of lymphocytes is approximately six-fold, and yet the differential blood count [e.g., which compares proportions of lymphocytes to total WBCs] is not noticeably changed. The cause according to Pischinger (1957) was that "up to now an important moment has been ignored, namely, the using up of white blood cells by the blood."

2. Leukocytolysis

As Pischinger stated in 1983:

> The question of the fate of the leukocytes (white blood cells) has occupied hematologists for a long time, especially since Undritz described leukocytes in smears of the fresh blood of healthy human beings and animals; he interpreted these leukocytes as breakdown cells (*Abbauzelle*). These vary in size (7 to 24 microns). The cytoplasm is uniformly basophilic, and only occasionally has an oxyphilic halo around the cell nucleus. The nucleus is completely structureless, pyknotic, and mainly dark in color. Occasionally there is nuclear fragmentation. There are similar breakdown cells for all the leukocytes in the blood. However, they are very rare. Koch, Heilmeyer, Laves and others make similar statements (Pischinger 1957, 627–629)

Later authors exposed *in vitro* blood samples to the effect of a partial vacuum before making the smears (Schröder, et al, 1959). This caused various progressive forms of disintegration of leukocytes to appear. All these methods have to do with genuine *resistance tests* for the cells. In this regard, H. J. Schröder presents not only relevant literature on the subject, but also a large number of original investigations regarding the extent to which physical, chemical, pharmacological, and clinical factors can influence white blood cell resistance. Exposure to a partial vacuum produces forms that are swollen to various degrees and sizes, depending on individual susceptibility.

In 1957 in an extensive paper on the fate of the white blood cells (leukocytes) in the blood, I showed that disintegration forms are also found in smears made of blood samples which had not been subjected to *in vitro* stress. To a large extent, these forms are similar in every aspect to those that appear following exposure to a partial vacuum. However, I do not believe that such cell remnants can be identified with Undritz's breakdown cells.

The following section is concerned less with the appearance of leukocytolysis as such, and more with the problem of whether and what relationships exist between leukocyte breakdown in the blood and the matrix regulation of the organism; namely, whether the process can be regarded as a cell-milieu reaction.

In the previously mentioned paper regarding the fate of the leukocytes, I demonstrated that the absolute number of leukocytes in the blood in the rabbit ear decreases between arteries and veins by an average of 17% (probability of error $p < 0.001$). It took a precise analysis of smears of capillary blood, prepared with special precautions, to explain this phenomenon. I found breakdown products of all types of leukocytes in all stages, which had the same forms that Schröder describes after damage by partial vacuum: from bare nuclei to reticular and flat-shaped fragments, whose origin as cell nuclei could only be confirmed by a positive Feulgen reaction, which reacts only to nuclear material. I therefore have no doubt that white blood cells in the blood are constantly being used up. At this point in the discussion, the extent of "leukocytolysis," as I have termed this phenomenon, should be investigated. Before continuing, however, we will first describe the special blood smear technique which minimizes the chance of creating of artifacts for example, by crushing cells.

It is best to prepare the smears with the edge of a slide 26 mm wide. The edge is specially prepared. Its middle part is rubbed with abrasive paper (glued to a uniformly flat base) 10–20 mm wide until there is a gap of about 1/10 to 1/20 mm when the spreading edge lies on the unabraded edges of the smear slide. With an angle of 25º to 30º between the two glass slides, the blood droplet—not too large—is spread along at moderate speed. To avoid clumping the leukocytes at the edges of the film of blood, it is necessary to work quickly between the time the blood droplet is placed on the slide and the smear is made. Naturally, the blood should not extend past the edges of the abraded part. In a preparation made in this way, the red blood cells should be spread in a single layer, and the white blood cells (almost) evenly distributed.

Conventional blood smear equipment is not acceptable for the following reasons:

1. The smear is too thick because the usual angle of 45º is too great.
2. It takes too long to to spread the droplet. As a result, the bulk of the leukocytes are found at the edges.
3. Grinding the gap on the spreading edge of the slide is difficult—the gap which is needed to keep from crushing the cells.

The mechanical stress to the cells in the spread blood droplet is simply due to the viscosity of the plasma. In the above-mentioned paper, "Concerning the fate of the leukocytes," I described and analyzed preparations made in this way of blood taken from different areas of the circulation. Lysis forms of white blood cells (leukocytes) are present in every smear of this kind. Their number varies according to the part of the body from which the blood sample is taken as well as the condition of the person being examined.

According to observations which Kellner made regarding the reaction of white blood cells with the addition of vitamin C *in vitro*, the destruction of the cells is at first unremarkable, but then proceeds suddenly (personal communication, Kellner). In blood smears of small animals (guinea pigs, rats, and mice) there are pale colored, apparently swollen white blood cells, undoubtedly the first morphological signs of the beginning of lysis, which can also be recognized in dark field microscopy. All of these signs cannot simply be artifacts. Incidentally, since that time in 1942, I have emphasized that if an "artifact" is critically assessed, it can make a biological statement. For example, if in a smear, a *Feulgen-positive lysis form* is right next to a *completely intact lymphocyte,* this means that they both must have been in *different states*, since they both were exposed to the same mechanical stress.

Including the lysis forms of white blood cells in the differential blood count in order to establish the extent of leukocytolysis seemed a sensible and constructive thing to do. In apparently healthy people there are 5–7% lysis forms in the blood taken from the fingertip. Given the absolute number of lymphocytes as 5,000, this is approximately 300 lysis forms per cubic millimeter. If there is an even distribution throughout the body and there is a total blood volume of 5 liters, this means that there are always 1–2 billion lymphocytes in the process of disintegrating. At this time there are no direct ways to determine how many *actually disappear completely* at any given instant, since the length of time from the beginning of the lysis process to the complete disappearance of the cell is unknown. However, it is possible to calculate how many disintegrate per unit of time by the size of the lymphocyte influx, assuming that these only disappear from the blood through lysis. With 25% lymphocytes out of a total of 5,000 white blood cells in a single mm^3, there are 1,250 lymphocytes per mm^3; in a total blood volume of 5 liters, this equals 1,250,000,000 x 5 = 6,250,000,000 — 6.25 x 10^9, in other words, about 6 *billion* lymphocytes. According to information in the literature — I have not investigated this personally — there should be an inflow of about 6 times this amount, which equals a total of 36 billion lymph cells. Accordingly, 30 billion must *disappear from the blood* for their numbers to remain constant; 30 billion every 24 hours; that is, 1.25 billion per hour, or about 20 million per minute, or 0.3 million per second. This figure is only valid for lymphocytes. To find the total number of white blood cells that disappear, the number of lymphocytes, assumed to be 25% of the total, must be multiplied by 4, which gives an estimate of 1.2 million per second. Considering that the blood smear only shows the pre-stages, namely elements that are already affected, then 1–2 billion cells *in the stages of disintegration* does not seem to improbable when compared to the 1.2 million cells that *have already disappeared.*

Present-day theoretical or clinical medicine does not pay much attention to the facts just described.

The next question that arises is, What is the significance of these phenomena and processes for the organism? Possibly, they serve to maintain the composition of the blood. Consider the following: disintegration of white blood cell releases protein, amino acid products, polysaccharides, lipids, nucleic acids, all sorts of tissue hormones and physico-chemical, oxidoreductive and surface-active complexes (Pischinger 1957, 627–629). The ultraviolet spectrogram of methanol extracts of white blood cells, and incidentally also of fibroblasts, reveals, for example, a specific color band in the same wavelength as our monocyte factor from the blood. So it comes as no surprise that the previously described investigations by Kellner show that substances released from disintegrating fibroblasts are able to affect the milieu, and also show how they do this. The same must be valid for white blood cells. In the case of both leukocytes and fibroblasts, both cases pertain to normal cells (Laves, Biermann).

Systematic counts of the disintegration types in smears from many people under various circumstances show that the relative and absolute numbers per mm³ of (fingertip) blood can vary a great deal. Our next task must be to shed some light on why these variations occur and what they mean. However, before doing that, I must add something about my counting technique: for reasons of precision I have always drawn blood up to the 1.0 mark (dilution 1:10) of the mixing pipette, and counted all 9 millimeter squares in Bürker's counting chamber. By counting a sample ten times, this method gives a standard deviation of S= (rounded up) +6%. The standard deviation of the mean runs at S_1 = approximately 2%. Naturally, I always used the same combination of counting chamber and pipette in comparative counts. When necessary, I took double measurements with two sets of equipment. The result came from the mean value. Single values were only used when they did not significantly deviate from each other by more than than was allowed by the difference between the two counting chambers. Otherwise a recount was done.

The following tables concern the question of the behavior of leukocytes (differential blood count) and lysis forms when the organism is influenced with monocyte serum remnant factor "M," and with penicillin. The differential blood count values are listed in detail in order to show that the counts for all leukocytes do change. I am listing the differential blood count values in detail in order to show that after subcutaneous injections, changes in the counts for all white blood cells are set into motion. This is well known in hematology, and shown, especially by Likint and also by Storck, to occur as short-term reactions following radiation therapy, hydrotherapy, hydrothermotherapy, douching (Blitzgüsse), light and heat treatments, and so forth. It is already known that the relationships of the lysis types also change under these kinds of influences (references in Schröder 1959).

First, there is an example of of the effect of a milliliter of Elpimed® (Gebro—M 1:100). This corresponds to a dry quantity of about 100 micrograms (Tab. 1). As a control, the same subject received a 1 ml subcutaneous normal saline injection one week later. (Tab. 2) G.P., male, 15 years, 53 kg; hypertrophic tonsils.

After that, there are three examples taken from clinical practice, in which the known varying effects of monocyte factors and normal saline solution are further investigated (Tab. 3, 4, and 5).

There is final investigation with guinea pigs to check whether influences can be differentiated from one other (Tables 6, 7, and 8).

Three animals were used in the investigation. They all weighed approximately 500 g, and all were all a pure strain from same litter.

The difference between the values in Tables 1, 3, 5, and 6 on one hand and Tables 2 and 7 on the other is clear. The injection of *normal saline* in cases 2 and 7 produces *hardly any* changes worth mentioning in the leukocyte and lysis numbers. There are minor variations, but as long as they are under the margin of error, they can be explained by the fact *that it is only in relation to the osmotic state of tissue that a "normal" saline solution can be considered "normal."* Since *the chemistry of the tissue is very different from that of a saline solution,* injecting saline solution must modify the milieu enough to cause the minor variations.

In contrast, a subcutaneous injection of 1 ml of 1:100 active serum extract factor "M," diluted in normal saline (corresponding, as stated before, to a dry quantity of approximately 100 micrograms of active substance), results in an obvious *increase in the lysis forms* in both humans and guinea pigs. In the Table 1, for example, the number of lysis forms increases from 325 to 382, 550, 1051, and 2720 in 9 hours, followed by a drop to 646 the next morning (24-hour cell count). It is striking that at the same time there is a *decrease* in the number of lymphocytes.

Examples 3, 5, and 7 showed the differential blood counts after a subcutaneous injection of 1 ml active serum extract (diluted 1:100). In these counts as well as that of Table 8, in the hours following treatment there was a large increase in the number of lysis forms, and again as in example 1, a decrease in the lymphocyte count.

However in comparison, this reaction to the administration of active extract "M" is absent in example 4. The lysis forms decrease in number earlier, and the lymphocytes increase. Therefore this reaction is the opposite of that in 3, 5, and 6.

This inverse reaction is not unusual (incidentally, it also happens in other fields). When assessing the phenomena of white blood cells, both this inverse reaction and the findings after administering penicillin (Table 8) are of special interest. Case 4 was under the same influence as the previous examples, namely, active serum extract "M" administration; the number of lysis forms remained high, and the lymphocyte count increased. Therefore, the cause for this must be sought in the *state of the organism,* and in fact, in the effective range of (active) serum extract "M," which promotes the increase of monocytes. As a matter of fact, the subject in Table 4 had focal lesions that, in our experience, have an adverse effect on autonomic basic functions.

Example 8 also stands out. The animal died three days after a subcutaneous injection of 4000 IU Na-Penicillin. The "toxicity" of penicillin in guinea pigs is known, but there is no clear explanation for this. Damage to the intestinal flora is considered a possible cause. However, in our experiment it is notable that the number of lysis forms *decreased* significantly from 690 to 419 in 24 hours. At the same time there was also a significant decrease in lymphocytes and eosinophils. In the guinea pig, the way the toxicity of subcutaneous penicillin was expressed is by adversely influencing the regulation of the blood. At present, no more can be said. Whether a similar process occurs in humans would be worth investigating. In any case, F. Perger established that penicillin reduces the serum calcium level in humans, although there is no "toxic" effect. However, hypersensitivities to penicillin must be interpreted as *regulatory disturbances,* specifically in the sense of inhibiting regulation.

Table 1[1]

Date	Time	Total No.	N	E	M	L	BF
Jan. 5	9:10	6490	2596	260	390	2920	325
	9:15	Injection 1 ml Elpimed®(Gebro)					
	9:35	6390	2860	254	254	2610	282
	10:00	5490	2910	220	275	1535	550
	10:45	4780	1720	240	287	1482	1051
	17:15	9380	3845	565	375	1875	2720
Jan. 6	8:00	5380	2045	270	323	2100	646

[1.] N=Neutrophils, E=Eosinophils, M-Monocytes, L=Lymphocytes, BF=Number of white blood cell breakdown forms

Table 2

Date	Time	Total No.	N	E	M	L	BF
Jan. 12	8:30	4780	1770	286	286	1960	478
	8:37	Subcutaneous injection of 1 cc normal saline					
	8:57	5510	2315	220	275	2480	220
	9:22	4530	1721	408	317	1676	408
	10:07	4620	2217	277	416	1340	370
	16:37	6330	3165	253	443	1962	507
Jan. 13	7:30	5600	2130	336	392	2350	392

Table 3: M. K., female, 50 y.o. cephalic

Date	Time	Total No.	N	E	B	M	L	BF
Feb. 8	9:00	5200	2288	208	104	104	2288	208
	9:05	Subcutaneous injection 1 ml Elpimed®						
	10:00	5100	2400	408	0	102	1834	356
	12:00	5000	1650	300	0	300	1950	800

Table 4: H. P., female, autonomic dystonia, focal, extensive surgical scars

Date	Time	Total No.	N	E	B	M	L	BF
Mar. 3	9:00	3700	1665	74	37	74	1665	185
	9:05	Subcutaneous injection 1 ml Elpimed®						
	10:00	3950	2014	79	0	158	1620	79
	11:00	4500	1845	45	45	180	2205	180

TABLE 5: R. P, MALE, 50 Y.O. VITIUM

Date	Time	Total No.	N	E	B	M	L	BF
Jan. 29	8:20	4910	1720	245	0	440	2360	145
	8:25	Subcutaneous injection 1 ml Elpimed®						
	9:20	4420	2080	221	0	486	1500	133
	12:20	5470	2740	95	0	383	1740	512

TABLE 6: GUINEA PIG, MALE, ELPIMED®

Date	Time	Total No.	N	E	M	L	BF
Jan. 20	9:15	10728	1952	867	217	6825	867
	9:25	Subcutaneous injection 0.2/400g kg					
	10:55	9640	3376	771	771	3180	1542
	16:55	8006	1865	648	466	3730	1297
Jan. 21	8:30	7900	1658	632	475	4110	1025

TABLE 7: GUINEA PIG, MALE, NORMAL SALINE

Date	Time	Total No.	N	E	M	L	BF
Jan. 27	10	7670	1189	1880	153	4065	384
		Subcutaneous injection 0.2/400g kg					
	10	6220	1120	1896	62	2550	497
	11	7100	1562	2200	71	2800	497
	16	7300	1533	2260	146	2810	438
Jan. 28	9	7840	1255	2038	157	4000	392

TABLE 8: GUINEA PIG, MALE, SODIUM- PENICILLIN

Date	Time	Total No.	N	E	M	L	BF
Feb. 3	9	11439	2860	2630	343	4916	690
		Injection 4000 IU penicillin in 0.2 cc normal saline					
	10	12022	1800	2650	120	6730	722
	16	5478	1646	930	220	2080	602
Feb. 4	8	8267	4962	992	413	1490	410

5 to 6.2—major decrease in lymphocytes and lysis forms.

We cannot conclude our examination of leukytolysis—or leukolysis—without taking into account two phenomena that have been reported in the literature. These are cytolysis by Freund-Kaminer, and the feasibility of tests using Pichlmayer's antilymphocyte or antigranulocyte serum, reported in Schroeder et al, 1959.

First of all, let us briefly touch on Schroeder's investigations and the associated previous literature, because they are informally connected with the above-mentioned findings.

Schroeder's preparation concerns a standard osmotic stressing of the blood in vitro, which shows a variety of lysis forms, depending on the susceptibility of the leukocytes. Other authors used different treatments, for example, uric acid and salt with a small addition of oxalate (Achard and Feuillie, 1907). Originally, smears of fresh animal and human blood were studied (Undritz 1941), using approximately the same techniques as those in the analyses described here. The same destruction forms are found in fresh (untreated) and treated blood.

Of the many phenomena that have been reported, certain ones deserve to be emphasized in connection with existing studies: According to investigations with all the above-mentioned methods, *the number of breakdown cells are influenced by drugs, and they are dose-dependent.* However, from his many studies using untreated blood, Koch's conclusions appear to be the most important ones. His opinion is that in the various reactions, the matter at hand is "not a direct effect of substances," but rather, concerns the response of breakdown cells as a *non-specific* reaction. This conclusion is supported by the fact that there are many different varieties of procedures associated with an increase in the number of breakdown cells. Koch speculates that various causes might result in the formation of a specific endogenous active compound in *every* situation in which the breakdown cell numbers increase, and this conclusion leads us to consider the results of the research at hand. According to my experience, it could be a matter of the substances in the active serum extract. These are present in varying amounts and normally *increase the lysis count.* However, there obviously must be other factors that inhibit leukocytolysis, such as was the case with penicillin in guinea pigs. Additionally, I do not believe that specific substances are involved. What does have this effect are the physical and physico-chemical activities that are unique to respective substances. This would mean that we are dealing with non-specific processes.

The observations made by J. Schröeder (1970) on the "Resistance test as an in-vitro method of determining the effects of antilymphocytic and anigranulocytic sera" (according to Pichlmayr and co-workers) deserve special attention; since this how the non-specific regulation and the immunosuppressive lymphocyte reactions are connected; these reactions at present are the deciding factors in matching donors and recipients for transplantation. Using original Pichlmayr antisera, Schröeder showed that the measuring the resistance of the white blood cells can test the efficacy and strength of antilymphocytic and antigranulocytic sera. If this is so, the antilymphocytic and antigranulocytic reactions must also, as it were, be rooted in the non-specific process. In the general scheme of things, this is understandable, *since when the basis of the life functions fails, the specific processes decline as well.*

To further understand the process of cell lysis, we must examine the cytolytic reaction in connection to leukocytolysis, according to Freund and Kaminer (1925).

It is well known that Freund and his co-worker Kaminer discovered that the serum of cancer patients has lost the ability, which healthy patients have, of breaking down cancer cells. Later it was shown that liver cells can be used for this test. Unfortunately, because there are other serious diseases that also show a loss of cytolytic capacity, the hope that a test for detecting cancer had been discovered was not realized. For this reason, the usefulness of this method to diagnose tumors was subsequently assessed with more caution. However, the phenomenon as such remained as a continued subject of research. In normal serum there are cytolysins, whose effects are inhibited in people who are ill. Stern and Willheim make a claim that is important for the examples at hand; namely that when cytolytic ability is absent, the RES (reticulo-endothelial system) is also damaged. They believe that one of the tasks of the RES is to guarantee the normal lytic behavior of the serum; since loading the RES with dye causes the loss of the lytic function of the serum (rabbit experiments).

In connection with these research results, we must refer back to the works of G. Kellner, previously described in detail, regarding the cell-milieu reaction in cell cultures that had alterations in the chemical and physical composition of the culture liquid. It was shown that if the properties of the liquid were not properly adjusted, there is massive cytolysis, and only at this point, when the milieu has obviously been adjusted by materials released from the cells, does the culture begin to grow again. Kellner also thoroughly examined these experiments physicochemically and saw that in this regard, the composition of the culture medium is the determining factor for the way the cells react; when the deviations are too great, cytolysis begins.

Leukocytolysis can be considered to be a process parallel to cytolysis in serum and cultures, so it appears to be dependent on physico-chemical factors such as pH, rH, and surface and boundary layer activity.

The fact that the suppression of leukocytolysis together with a simultaneous decrease in lymphocytes causes the death of the animal, confirms that *leukocytolysis is one of the elementary process of the organism.* Leukocytolysis is clearly triggered by a physico-chemical and/or an energetically based change in the milieu, in this case, in the blood. In this sense—to put it into general terms—one can speak of a regulatory cell-environment reaction for the loose basic (connective) tissue, which we described earlier. Here, as in that example, the milieu (in this case, the blood plasma) must be restored to an adequate condition through the destroyed—note: *not* degenerated—cells. It should not be overlooked that the quantities of material released by cell breakdown are not small. Kellner (personal communication) estimated that at the instant of counting, up to a half gram of leukocytes by (liquid) weight was disintegrating and their substances were having an effect.

This leads to a further series of similar questions, such as, as we already asked, how many leukocytes actually degenerate? Is the quantity of leukocytes parallel to the number of lysis forms, or is it the other way around? That is, do lysis forms decrease because their complete disintegration takes place more rapidly, or do the leukocytes become more resistant? We know from Koch that number of breakdown cells parallels the serum peroxydase content, which comes from the granulocytes. According to Koch, the number of breakdown cells could be used as an indicator for intravital leukocyte disintegration. A further question concerns the fate and the function of the certainly not insubstantial quantity of nucleic acids that that are constantly being released during lysis. A third question must also be considered, namely, where do the substances come from that increase the number of monocytes? These substances, it appears, are equally as important as substances that stimulate leukocytolysis. Finally, we cannot ignore the cytolytic processes of the leukocytes in blood diseases, in which the breakdown forms are particularly abundant. These questions must be the object of special hematological work.

Conclusion

Leukocytolysis in and of itself is not important, but its contribution to the extracellular matrix is. This assumes the readiness ('sensitivity') of a given percentage of leukocytes, corresponding to the individual state, to disintegrate. The reactivity of the extracellular matrix can be estimated by considering this "pool" (Heine 1988c; Draczynski 1998a, b; article by Perger, 1990).

2.1. Regulation of the Tumor Extracellular Matrix

The fibroblast is the regulatory center of the normal extracellular matrix. In contrast to this, in fast-growing, malignant tumors, every cell is able to synthesize tumor-specific extracellular matrix by shedding "tumor matrix vesicles" (Fig. 8, Heine 1987a, 1988a). This high-energy reactive situation supersedes normal matrix regulation.

Through paracrine and autocrine stimulation, the tumor process spreads peripherally and progressively includes adjacent tissue (Fig. 8, 9). The normal extracellular matrix is destroyed by proteolytic and hydrolytic enzymes from the disintegrated tumor matrix vesicles. Of these enzymes, plasmin, in particular, breaks down proteoglycans, and tumor matrix vesicles release the plasminogen activators that transform plasminogen into the plasmin (Heine 1987a). Correspondingly in malignancies, the polysaccharide complexes show an "isolation" of the extracellular matrix. With increasing de-differentiation and malignancy, this leads to a preponderance of hyaluronic acid (a highly negatively-charged, elongated glucosaminoglycan that exhibits no protein binding) (Heine 1987a). It is worth noting here, that hyaluronic acid is an evolutionarily primitive part of the extracellular matrix of all multicellular organisms. In human beings beings it appears at the end of the second week of development, when the mesenchyme develops [at an embryological phase in which it is necessary] (Fig. 7). Only later do proteoglycans form. The importance of this polyanion lies in the fact that it stimulates mitosis and at the same time inhibits differentiation (Toole 1983, Heine 1987a). In malignancies, this phenomenon appears in the wrong place at the wrong time (1987a). Radiation and surgery are important for reducing the size of the tumor mass. Since these measures as well as intravenous cytostatic drugs have the additional effect of damaging the intact extracellular matrix, adjuvant treatment (which prevents further tumor formation) is recommended to activate the basic regulation of healthy tissue and to stimulate the regulation of the extracellular matrix before, during, and after traditional medical treatment. All of the so-called natural therapies are appropriate for this, since they stimulate the capacities and the reserves the organism has for self-healing. Even during illnesses such potentials and reserves are usually available and can be therapeutically stimulated (cf Leupold 1945, 1954).

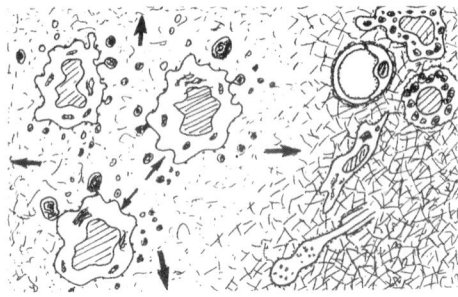

Figure 8. Regulation of the tumor extracellular matrix by the tumor vesicle. Three tumor cells shedding matrix vesicles are depicted in the left half of the picture. The extracellular matrix has fallen into pieces. The arrows show the spread of this disorder into the immediate environment as well as the reciprocal influence of the tumor cells on each other. The right side of the picture shows a normal interlinked ground substance with typical cell components (*see* Fig. 8)

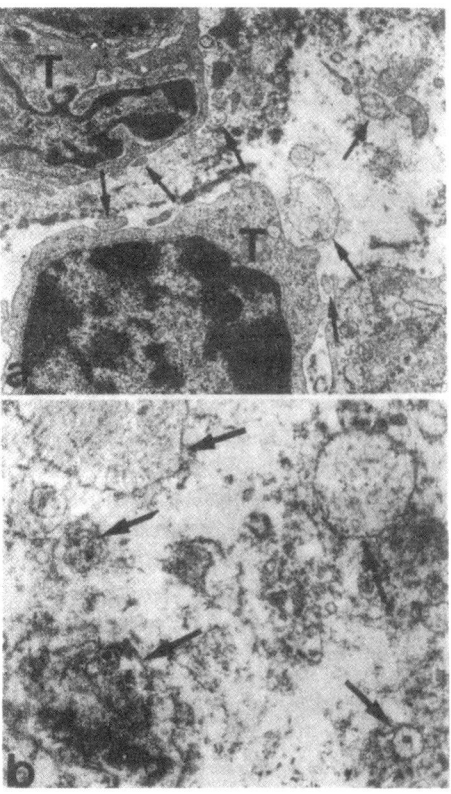

Figure 9. Scirrhous breast carcinoma. a) Formation of tumor matrix vesicles (arrows) on the surface of tumor cells. x 5,000. b) Degenerating tumor matrix vesicles in the tumor extracellular matrix (arrows). x 7,500.

3. The Significance of Chronobiology

The importance of social contacts, an individual life rhythm, and mental, dietary and climatic factors, in maintaining and activating matrix regulation should not be underestimated. This indicates a close connection between exogenous and endogenous rhythms. Since energetically open systems are compelled to fluctuate because of their instability—which is related to the principle of facilitation and inhibition—the rhythms on a molecular level are open to the larger rhythms of a higher order, such as the rhythm of night and day or the rhythm of the seasons. The most important intracellular "pacemaker" is the rhythmic synthesis of ATP by the cell mitochondria (Priebe 1981). The pacemaker of the cell membrane is the circadian rhythm of the "secondary messenger" associated with the cell membrane, as is the case with the sympathetic-associated cyclic adenosine monophosphate (cAMP) (Lemmer 1983).

Finally, in the extracellular space, the pacemaker is the rhythm in the relationship of sugar bipolymers to the molecular swarms of fluid/crystal water and ions. These relationships can be recognized, as seen in the higher control the circadian rhythm has over urine excretion. As early as 1893, Qinque observed an inverse rhythm in patients urination. At the present time, nocturnal polyuria (frequent nighttime urination) is still an important symptom in patients with heart failure. Since the sugar bipolymers in the extracellular matrix are capable of binding with water, nocturnal polyuria is an indication, among other things, of a disturbed extracellular matrix which is less able to bind with water. Heine (1990 a, b) observed a circadian rhythm for the proteoglycans and glycosaminoglycans of the extracellular matrix; at night there was a surge and an increase in sulfation, during the day a higher water- and protein-bonding. The sugar biopolymers were decreased at night in cases of tumors and chronic illnesses. This being the case, this decrease in sugar biopolymers can blur or reverse the circadian rhythm of the extracellular matrix if there is a progressive illness. This also depends on individual's predisposition. Chemotherapeutic measures appear to be especially destructive to the circadian rhythm of the extracellular matrix (Heine 1990b).

Circadian temperature variations also point out molecular rhythms in the relationship between the extracellular matrix and the cells. This is made clearer by investigations carried out by Gaultherie and Gros (1977) of breast cancer. In 26 women, after a precise grading and staging of the tumor, as well as synchronization as regards activity, sleeping time, mealtimes and room temperature, the skin temperatures of the affected and the non-affected breasts were measured continuously by telemetry for nine days. In 11 patients with a slow-growing, well-differentiated carcinoma, analysis of the findings also showed a circadian rhythm in the temperature on the affected side, but the amplitude was reduced and the temperature maximum appeared earlier. In 15 patients with rapidly growing tumors with histologically undifferentiated cells, no circadian rhythm (only a chaotic rhythm) was observed in the skin temperature on the affected side.

This finding agrees with the previously mentioned "de-differentiation" of the extracellular matrix, which contains a predominance of hyaluronic acid (see page 30). As in developmental history, hyaluronic acid initially predominates in the extracellular matrix, so rhythms are also subject to a maturing process. An example of this, which only appears after the 7th year of life and still can be observed in adulthood, is the circadian rhythm in body temperature with regard to amplitude and phase (Abe et al., 1978). Anthroposophic medicine presented convincing findings for a 7–year rhythm in mental and spiritual development in connection with the body (Fintelmann 1988). Conscious observation of the relationship between the extracellular matrix and the cell is made possible by the sensation of pain, which is also subject to a circadian rhythm. Disturbances in health that are not clearly delineated, endogenous depression, and diffuse pains of psychogenic type indicate disturbances of the molecular rhythm in the relationships between the cell and the extracellular matrix. Among other things, they appear in disturbances in the rhythm of sleeping to waking (Summaries in Lemmer 1983 and Hildebrandt 1987).

3.1. Psychosomatic Stress Reaction Processes

Disturbed rhythms are expressions of stress, and moreover, extended stress leads to disturbance of all biological rhythms. As first described by Selye (1971), an organism reacts to stress with a characteristic and unspecific reaction, the "general adaptation syndrome" (cf article, Perger 1990).

Following that, there is an "alarm phase of general activation" ("sympathetic shock phase" according to Perger 1990), and a "phase of resistance" ("parasympathetic counter-shock phase" according to Perger). When coping with stressors, the system usually settles into a seven-day rhythm. In other cases, the organism switches to a phase of exhaustion of the hypothalamus-pituitary-adrenal axis, resulting in general injury to all organs. In addition, Selye recognized the central function of corticosteroids in the response to stress, and regarded them as pathogenic agents. He subsequently defined stress as the "unspecific reaction of the body to demands of any kind" (Selye 1971). In order to distinguish between pathogenic and non-harmful stress, in 1974 he introduced the concepts "eustress" and "distress." Unfortunately, this distinction is not in common usage. The medical definition of stress is "strong bodily loading, which can lead to bodily injury."

3.1.1. Controllable and Uncontrollable Stress

Stressors are always comprised of psychological and physiological stimuli. Thus the focus is always on the importance of subjective evaluation of situational demands (Lazarus and Folkman 1984).

These authors distinguish between a *first evaluation*, in which the individual evaluates an event by the effect on his own person, a *second evaluation* as an assessment of the coping possibilities, and a third phase, a closing *reevaluation* of the situation. The *coping process* can proceed from this point.

Ursin and Olff (1992) stress the need to distinguish between two types of stress-related activation: *phasic activation* appears with successful coping and is accompanied by an increased release of adrenalin, quicker pulse, and a marked increase of plasma testosterone. In the absence of coping, this turn into *tonic activation* overflowing with a higher cortisol level, from which, after a time, psychosomatic disturbances can develop (summary by Huether et al. 1997).

Huether et al. (1997) assume a stress-reaction process (SRP) because of the partly contradictory meanings of the concept of stress (Fig. 10).

If an individual manages to end a SRP on his own, this activity is called a *controllable stress-reaction process*. Otherwise it is termed an *uncontrollable stress-reaction process*. Both forms of SRP are necessary for a normal, healthy development. Too much either controllable SRP or uncontrollable SRP can cause developmental errors and illnesses. An SRP is characterized by a continuous reaction and feedback between the trigger of the stress and a multiphasic cognitive and emotional assessment of the stress. This is primarily with respect to their importance for the stability and integrity of the individual as well as the physiological and behavior-related reactions to the act of coping Huether et al. 1997).

For dealing with the controllable SRP, the individual has coping strategies which require a certain amount of exertion and which are not yet completely a matter of course. With repetition, this approaches appropriate adaptation ratios as the efficiency of the associated behavioral strategies improves. Repeated, controllable stress loads eventually lead to a successive facilitation, stabilization and improved efficiency of reactivity.

According to studies performed on young research animals, there are increased numbers of neuron connections in the neuronal regions involved because of increased dendritic branching of the pyramidal cells, proliferation of glial cells, and a greater concentration of synaptic terminals (Greenough and Bailey 1988). Adult animals have decreased anxiety in new environments and a decreased cortisol secretions in a variety of stress situations (Levine et al. 1967, Alana et al. 1986). Controllable stress can successfully stabilize evaluation and coping patterns of an individual and can make them permanently available. Thus, in altered living conditions, controllable stress is of great importance for the progressive adaptation of an individual (summary Huether et al. 1997).

In contrast, uncontrollable stress denotes a process that the affected individual cannot cope with on his own, because the evaluation and coping strategies at his disposal are not practical or do not apply to the situation.

Behavioral biological and neurobiological findings show that neuroendocrine processes that arise as a result of uncontrollable SRP lead to predominately degenerative neuronal connections through plastic changes. Through this, in the face of new challenges, behavioral patterns that are no longer useful are destabilized and dissolved. It has long been recognized that in this kind of destabilization process there is always the risk that the decompensation of the affected individual will cause stress-induced illnesses (e.g., arteriosclerosis, coronary heart conditions, myocardial infarction). At the same time this implies that there is also a constructive possibility—that of separating oneself from behavioral patterns that are no longer useful and developing new and more effective ones in their place. Only this can generate basic changes in thinking, feeling, and interacting, which are important requirements in human development (hypothetically, the animal experimental findings can thus be transferred to humans, which speaks in its favor; Willner 1984).

This flexibility in conjunction with new, challenging situations is especially needed during transitional stages of life such as puberty or midlife crisis. A person who is able to turn both controllable-SRP and uncontrollable-SRP into a flexible pattern of behavior, will find it easier to cope with difficult life situations. According to Huether et al. (1997), this person possesses a "protective factor" that protects him from a continuous activation of his neuroendocrine stress reaction, which, in other people, can lead to immunosuppression and a higher susceptibility to diseases.

It is important for the nursing infant to acquire these protective factors, since brain development is at a crucial stage during which neural circuits are being laid down that create a lifelong basis for the evaluation and coping system of the individual. In uncontrollable stress-loading, this does not occur.

"Optimal frustration" therefore seems to be necessary for healthy psychological development; in contrast, uncontrollable stress is clearly a trigger for a structural reorganization process in the brain. Optimal stress, as a novel and heavily weighted demand, can actually help the adaptation of the individual since he cannot deal with the stress by using his previously developed repertoire of behavioral and coping strategies. An "optimal stress factor" presents the best foundation for healthy human development and works as a protective factor for psychologically, neurologically, and immunologically mediated diseases (Huether et al. 1997).

3.1.2. Body Awareness and Stress Management

Three abilities that direct the stress reaction—goal-oriented thinking, decision-making, and body awareness—are localized in the anterior forehead area of the brain (prefrontal cortex). Body awareness is in the background of all cognitive functions, and accompanies our mental images, whether new or remembered, and perceives them as comfortable or uncomfortable.

In humans, the ability to associate body awarenesses (feelings, "somatic marker" Damasio 1995) with cognition is partly inborn; in some measure it is developed as part of the socialization of the individual (i.e., coping ability).

Just how important these types of somatic markers are can be clearly noted in the behavior of a brain-injured patients. Their ability to perceive feelings has been largely lost, and along with that, the ability to make decisions has also disappeared (Damasio 1995).

Not only are feelings the basis of our decisions; they help even more in thinking, because they help us make preliminary decisions and unconsciously urge us in a set direction. Feelings also warn us of things that we have had prior negative experiences with, or guide our attention to something important. This guides the stress management. In evolutionary history, the body was present first. After that, the simpler abilities of the brain developed, such as the awareness of the state of the body. Finally, the more complex abilities such as abstract thinking and self-awareness appeared. This is reflected in our thinking. The more recently developed abilities have indeed achieved a certain measure of autonomy, but they are permeated with the earlier evolutionary structures that represent the biological survival interests of the organism.

3.1.3. Neurological Basis of the Stress-Reaction Process

It is impossible for a unidirectional model to represent the highly complex, interacting neuroendocrine regulatory circuits. At present, the stress phenomenon in its biological process can be simplified in the following manner: at the beginning of the process, stimulus patterns which are novel and/or perceived as threatening are registered by the brain as emotional recognition. With this, a particularly high neuron activity appears in the prefrontal cortex. This leads to characteristic activity patterns in the limbic system, which connects the frontal cortex and the combined neocortex with the archicortex (hippocampus) and the paleocortex (olfactory brain). In both of these evolutionarily primitive parts of the brain, the emotionality, drive, instinctive reactions, as well as motivation, connect with each other and are assessed.

The amygdala is of special significance in this, because it is here that the incoming information receives its "affective coloration" (Tönung). The limbic system is also the overriding control center for the endocrine and autonomic nuclei in the hypothalamus and brainstem. The regions of noradrenergic nuclei are on this level. On one hand, these have a positive feedback to the cortex, limbic system, and the hypothalamus. On the other hand, they stimulate the peripheral sympathetic and adrenomedullary system. The lowest level of SRP then produces a rise in the catecholamine level in the blood and which affects the peripheral organs. It is true that catecholamines cannot cross the blood-brain barrier, but a series of indirect effects unfold, which range from the change in blood circulation in the brain to the modification of the supply of substrates for brain metabolism (Fig. 10; summary Huether et al. 1997).

The positive feedback between the regions of noradrenergic nuclei of the brainstem and the levels above them would lead to ever stronger swings in activity, if it were not for simultaneous activation of the hypothalamus and with that stimulation of the hypothalamus-pituitary-adrenal axis. This leads to a subsequent rise in the cortisol level in the blood, which leads to the opposite regulation, that is, inhibition. It is important that cortisol can pass through the blood-brain barrier and can directly influence brain cell receptors and their functions. This keeps the activation processes triggered by catecholamines under control (Munck et al. 1984). Results of animal experiments show that over an extended time, a continuous increase in the cortisol level has a degenerative effect on the CNS: the noradrenalin release is decreased, the cerebral energy exchange is restricted, and the production of neurotropic factors is suppressed. This may lead to a degeneration in noradrenergic axons in the cortex, and to the die-off of pyramidal cells in the hippocampus. Finally, high cortisol levels facilitate the dissolution of all such learned behavioral reactions that are appropriate in successfully ending the stress-reaction processes (Uno et al. 1989, Bohus and de Wied 1980, summary Huether et al. 1997).

Inadequate coping with stress increases the latent inflammatory readiness of the loose connective tissue. This promotes allergic processes, autoimmune processes, and chronic illnesses, since futile behavioral patterns promote the phenomenon of neurogenic inflammation. The term "neurogenic inflammation" implies that the autonomic sensory axons, which end in the extracellular matrix of the connective tissue, not only have afferent but also efferent functions. Under prolonged stress, these axons can release inflammatory mediators such as interleukin-1, substance P, catecholamines, etc. Thus, in the immediate vicinity that these substances are released, the neurotropic mast cells are stimulated to release mediators from their granules. Of these mediators, histamin, serotonin, prostaglandins, and leukotrines trigger a superfine focus of inflammation. (summary Weihe 1990). This focus is remodeled by fibroblasts through highly structured collagen type I (Heine 1995, 1997a). The collagen sleeves thus formed, which surround the terminal autonomic sensory axons, can only be observed at very high magnification. The information that these sleeves present to the affected axons is decoded as pain by the brain. This self-initiated positive feedback encourages the further formation of collagen sleeves around neighboring axons. The process spreads and leads to symptomology of fibromyalgia. Since this is an illness with an enhanced reaction to a previously encountered antigen, this always leads us back to find a stress reaction that cannot be coped with (summary by Heine 1997a).

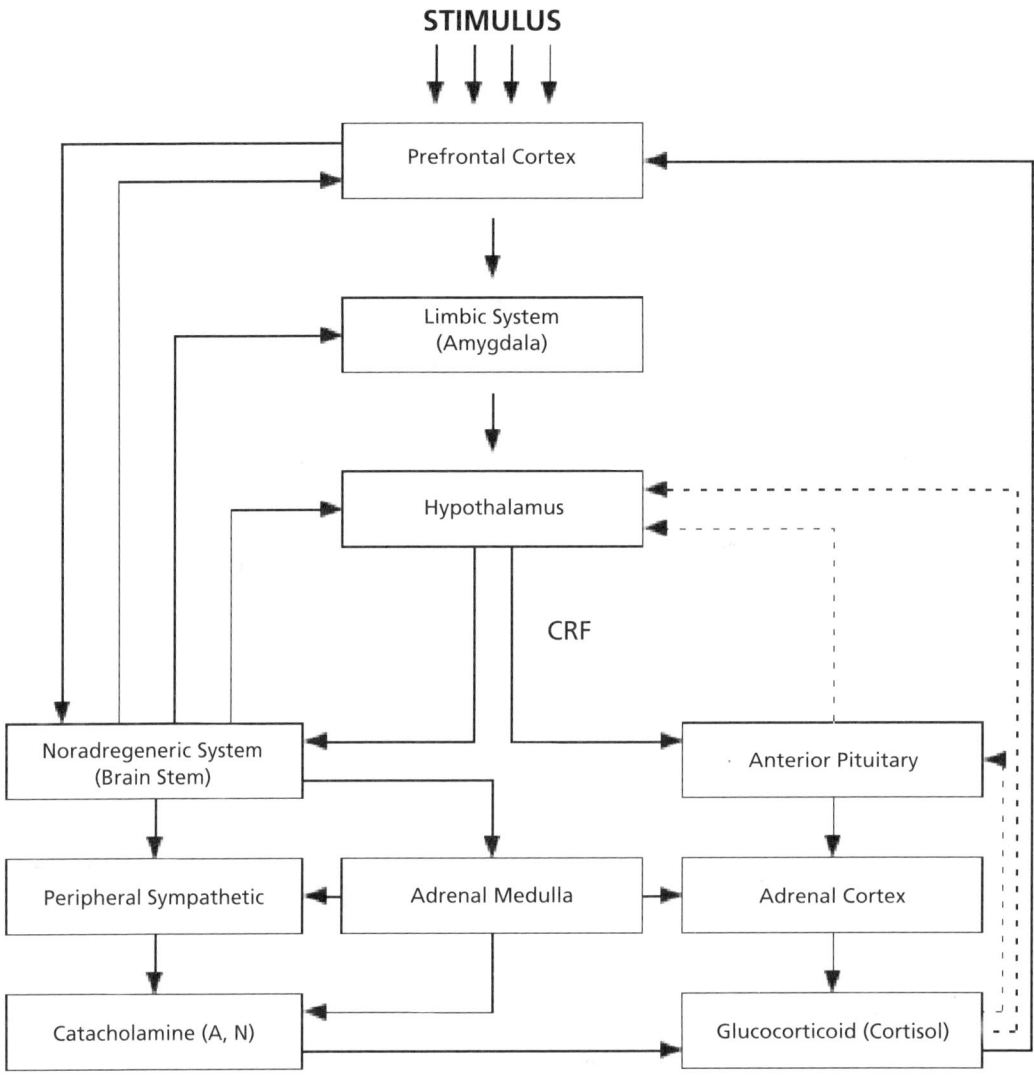

Figure 10: Summary of the most important participants in the stress-reaction process (→Activation, →Suppression), CRF = corticotropin releasing factor, ACTH = adrenocorticortotropic hormone (after Huether et al. 1997).

4. Topography of Extracellular Matrix Distribution

Loose, "soft" connective tissue is particularly rich in extracellular matrix, and so is particularly reactive. It retains its complete resemblance to the embryonic mesenchyme. Loose connective tissue has a typical distribution pattern: it accompanies all capillaries, forms the reticular cell tissue under epithelial cell groups such as the epidermis and mucous membranes (e.g., the tunica propria of the esophagus and the gastrointestinal tract), the splenic pulp, lymphatic tissue, fat tissue, the uterine mucosa, the ovarian cortex, the tooth pulp, and the Virchow-Robin spaces in the central nervous system. As loose endoneurium, endomysium, and peritendineum internum, connective tissue accompanies the vessels between nerve, muscle, and tendon fibers. It forms the interstitial connective tissue. This becomes the interlobular connective tissue which bears vessels and nerves and which divides glandular tissue into lobules. In addition, the adventitia of the larger vessels, serous membrane tissues, endocardium, soft brain membranes, interstitial tissues of the lung periphery, synovial membranes of the joint capsules, and the innermost layer of periosteum (cambium), all consist of loose connective tissue.

The finely structured connective tissue of the organ capsules forms a transition to firm and hard connective tissue types. Tendons, fascia, ligaments, aponeuroses, dura mater, the stratum reticulare of the skin, the cornea, cartilage, bone, and dentin already have the characteristics of organs. Even these "organs" are provided with loose connective tissue that accompanies vessels and nerves (e.g., Haversian and Volkmann's canals in bone, and dentin canaliculi in teeth). Only the cornea and postnatal joint cartilages are free of blood vessels.

The connection between hard substances and the soft connective tissue that penetrates them, accompanied by vessels and nerves, clearly illustrates of the importance of the loose connective tissue that contains the extracellular matrix. Its purpose, which also includes maintaining the functional capacity of the hard substance, is obvious here.

Recently, the closely circumscribed perforations (diameter 3–7 mm) in the superficial fascia of the body have been shown to be particularly significant, as only there, ensheathed in loose connective tissue, can the blood vessel and nerve bundles of the skin penetrate deeply. These are morphological correlates to acupuncture points (Heine 1988b). After connection with the larger vessels and nerves, the latter finally reach the spinal nerves of the CNS. Analogously, fine vessel and nerve bundles, enclosed by loose connective tissue, leave the dura mater in the area of ossified cranial sutures through meandering channels in the bones and appear in the skin of the scalp. The exit points in the bone also correspond to acupuncture points (Heine 1988b). Since these "points" (or rather, perforations) are always found in the same places on people regardless of racial differences, the points seem to be genetic in origin.

Obviously, the acupuncture points can develop discrete clinical disease patterns. (This could be a causal explanation for peripheral neuropathies, Sudeck's syndrome, painful triggerpoints, etc.) These findings also offer a rational basis for neural therapy and related methods (*see* contribution by Bergsmnn S. 91).

5. Structural Components of the Extracellular Matrix

5.1. Glycosaminoglycans (GAGs)

Glycosaminoglycans (GAGs) present unbranched, negatively-charged, linear carbohydrate chains with characteristic repeated disaccharide units of hexosamines (D glucosamine, D-glactosamine) and uronic acids (D-glucuronic acid, L-iduronic acid). The GAGs that are the most important for the extracellular matrix are: hyaluronic acid, chondroitin sulphate, dermatin sulphate, keratin sulphate, heparan sulphate and heparin. Hyaluronic acid is the largest and the most biologically important GAG, with a molecular weight of about 100,000 to several million Daltons, and consists of disaccharide polymers of N-acetylglucosamine and D-glucouronic acid. Only hyaluronic acid and heparin appear as free, non-protein-bound GAGs, and are correspondingly water-soluble. Among the GAGS, keratan sulphate is an exception, since in it the uronic acid is replaced by galactose. The glucosamine remnants of heparan sulphate and heparin contain N-sulphated groups (Matthews 1975, Hascall and Hascall 1991).

Sulfate and carboxyl groups give the GAGs a strongly negative charge. Because their charges repel each other, the polymers are stretched out, and this even determines their biological qualities and interactions with other molecules. In evolutionary terms, hyaluronic acid is the oldest carbohydrate biopolymer of multicellular life forms, and can be demonstrated as far back as the ground substance of sponges (Matthews 1975). Hyaluronic acid is noted for being able to bind to proteoglycans, and this is the starting point both in the evolutionary and developmental history of multicelled organisms (Matthews 1975).

Thus, in the developmental stages of vertebrates, hyaluronic acid is also the first component of the extracellular matrix to appear in the embryo. It is important because it regulates cell division and cell migration, as well as inhibiting premature cell specialization (its significance in tumor events has already been pointed out). Due to its high negative charge, hyaluronic acid is particularly capable of binding water and of ion exchange. Compared to other biopolymers, hyaluronic acid in aqueous solution can accept a very large domain (spread-out area), measured in molecular weight. The water requirement for full extension is so large that even at a concentration of 0.1% (W/N), the individual molecule already takes on a zigzag form and begins to overlap itself.

GAG and proteoglycan domains of this kind have the ability to exclude molecules of a certain size. They thus form a primitive first defense system. The domains are elastically deformable, and thus offer a shock-absorbing, viscoelastic system with an energy-absorbing effect. Together with proteoglycans, hyaluronic acid forms an important lubricant, and is the main component of the synovial membranes. The common denominator in all of the combined rheumatic conditions, no matter what their varied etiology, can be found in the pathological changes in the hyaluronic acid and the proteoglycans bound to it (Heine and Schaeg 1979, summary by Hascall and Hascall 1991).

Heparin is a GAG of special physiological importance. As the only sugar biopolymer, it is stored in a specific form of mast cell, that is, in the membrane-enclosed vesicle of the basophilic granulocyte, and is released as needed. It is the author's opinion that, not only can mast cells be basophilic granulocytes, which after being formed in the bone marrow migrate out of the bloodstream; mast cells can also arise locally from primitive mesenchymal cells, or they can correspond to the reticular cells of the bone marrow. From personal observations, fibrocytes and local macrophages (histiocytes) have the capacity to take up large quantities of mast cell granules as well as heparin that has been injected transepidermally or subcutaneously (Heine 1984, 1986).

Mast cells are capable of ameboid movement and are neurotropic, i.e., they travel towards the termini of autonomic axons. Since mast cells usually make their way to the immediate vicinity of capillaries and the adventitia of larger vessels, this is where they are mainly observed. One significant stimulus for degranulation comes from catecholamines that are released into the extracellular matrix, for example, under conditions of increased stress. However, changes of pH towards acidity, the binding of certain antigenic substances (i.e., "allergens") via the IgE receptors of the mast cell surfaces, inflammatory mediators (e.g., certain leukotrienes as "slow reacting substance of anaphylaxis"), etc., lead to mast cell degranulation (Heine and Schaeg 1979, König 1983). It is a common mistake to focus mainly on the coagulation-inhibiting function of heparin. Heparin takes part in all the regulatory processes of the extracellular matrix. Heparin is involved in regulation of the lipolysis of circulating lipoproteins by activating lipoprotein lipase in the endothelium; heparin promotes the aggregation of lymphoid cells (*Lymphzellen*), and activates the protein kinase of muscle cells. Heparin promotes the synthesis of extracellular matrix by fibrocytes, takes part in the activation of about 50 enzymes, and is involved in the translation and transcription mechanisms of DNA and RNA. Heparin promotes the breakdown of anaphylactic complement factor C3a, inhibits the effects of interferon, regulates thrombospondin synthesis in endothelial and smooth muscle cells, and is involved in collagen synthesis and collagen fiber polymerization. In addition, it is well documented that heparin modulates circulating growth factors, has a significant role in the control of plasminogen activators, and is involved in bonding with angiogenesis factors, histadine-rich glycoproteins, fibronectin, and platelet factor 4 (Summary Engelbrecht 1977).

6. Sugars of the Cell Surface: The Glycocalyx

The sugar surface film of the cell mediates functionally between the cell interior and the extracellular space (Fig. 1). It makes up the cell-specific and organ-specific receptor coating of a cell, which significantly determines the function and integrity of the cell. The glycocalyx sugars consist of branched oligosaccharides with terminal N-acetylneuraminic acid, and take root in the proteins and lipids (glycoproteins and glycolipids) of the cell membrane. Individual glycocalyx components have long been recognized as blood group substances and transplant antigens. Loss of the terminal neuraminic acid can mean is that a cell is aging, and this causes the cell to be recognized and eliminated by the cells of the reticuloendothelial system. Analogously, on the inside of the cell membrane there are filamented glycoproteins that are related to spectrin, vinculin, and actomyosin. They attach to the integral proteins of the cell membrane as well as to the filaments of the cytoskeleton. In this way, important membrane functions are significantly controlled, such as fluidity of the membrane and thus its capacity for depolarization and repolarization, the activation of secondary messengers in the membrane, receptor properties, etc (summary Hay 1991). Due to its negative charge, the glycocalyx has its own electrical potential (z-*Potential*), which differs from the potentials of the extracellular matrix and the cell membrane. The glycocalyx reacts with a change in its own potential following a certain alteration in the charge of the extracellular matrix. This suffices to activate the system of membrane secondary messengers in all connective tissue cells, immune cells, and epithelial cells (particularly the adenylate cyclase system, cAMP system, cGMP system, inositol phosphate, G-proteins). In contrast, the membranes of nerve cells and their processes, the membranes of cardiac muscle cells, smooth muscle cells, and striated muscle fibers become depolarized. This then leads to a cell-specific response (e.g., appropriate synthesis of extracellular matrix, muscle contraction, or beginning of a nerve potential).

The glycocalyx has binding sites for GAGs (particularly hyaluronic acid, heparan sulfate, and chondroitin sulfate); in part, the GAGs are bound covalently to hydroxyaminic acid remnants of the membrane glycoproteins.

The glycocalyx GAGs are coupled to one another by bivalent cations (mainly calcium ions). The glycocalyx GAGs participate in growth control, mitosis frequency, and the active movement of cells. The glycocalyx GAGs of the outer cell surface also attach to proteoglycans and structural glycoproteins of the cell matrix. In this way, via the glycoproteins and lipids of the membrane, contact is made between the extracellular matrix in the extracellular space and the cytoskeleton inside the cell. There are indications that proteoglycans and certain interconnected proteins of the ground substance (e.g., fibronectin) can penetrate the cell membrane directly, and thus make immediate contact with the filamentous cytoskeleton (Yamada 1991, Iozzo 1985). In this way, an exceptionally precise and rapid extracellular to intracellular, or vice versa, information transfer would be possible (similar to light conduction along fiberglass fibers) (Heine 1986). In tumor cells, these relationships are significantly disturbed since the product of a certain oncogen (ras oncogen) masks the vinculin on the inner side of the cell membrane, so that the cytoskeleton can no longer adhere uniformly to the membrane. Thus, there is a severe alteration in all cell functions (Hay 1983), which, among other effects, favors the formation of tumor matrix vesicles (Heine 1987a).

Disturbances of the extracellular matrix can alter the glycocalyx sugars to such an extent that there are major changes in cell behavior. This releases sugar components after the loss of the terminal sialic

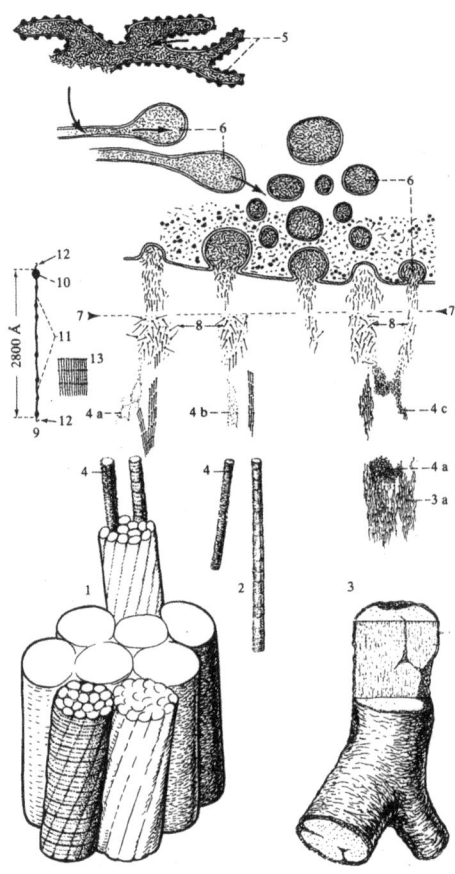

Figure 11. Diagram of formation of (1) collagen, (2) reticular fibers, (3) elastic fibers, and (4) PG/GAGs. By amino acids being taken up into the cytoplasm of the fibroblast, tropocollagen (tropoelastin) and proteoglycans are synthesized in the rough endoplasmic reticulum (5), and are completed in the smooth endoplasmic reticulum and extruded through the Golgi apparatus (6). The various types of fibers, apparently depending on local conditions, are oriented and formed in close proximity to the cells (7). This gives the PG/GAGs (4a, b, c) great quantitative and qualitative significance. (8) Tropocollagen molecule; (9) greatly enlarged tropocollagen molecule; head (10) and tail (11) with telopeptides (12). Fibrils of collagenous and reticular fibers are formed by the joining of tropocollagen molecules in a staggered arrangement, in which each molecule is a distance of one fourth of the molecular length from the next one. (13) "Long-spacing" collagen is formed by lateral aggregation of tropocollagen molecules with no staggered displacement. During the formation of elastic fibers (3), elastin-specific desmosin and isodesmosin are formed by the crosslinking of 3a and 4a.

acid. Bacteria, for example, can identify these sugar components by their carbohydrate chains and attach to them. However, plant, microbial, and the body's own lectins can also bind to the glycocalyx sugars and provoke a variety of cell reactions (division and synthesis capacity). The term "lectins" is understood to mean glycoconjugates and proteins that mainly have the function of recognizing sugar structures on cell membranes or in soluble glycoconjugates. At the same time, lectins have no correlation to the specific recognition mechanisms of the immune system, and they to not exhibit any enzymatic activity. Obviously they play a major role in so-called immune modulation (Uhlenbruck et al., 1986, Heine 1988). More important is the fact that there are kinds of lectins of the tumor cell glycocalyx that could be responsible for the spread of organotropic metastases. For example, the glycocalyx of liver cells has a sugar component that both forms a receptor for serum asialoglycoprotein and also can stimulate mitosis. It is assumed that certain lectins "are determining factors for the adhesion of circulating tumor cells" (Uhlenbruck et al. 1986). It is further assumed that these lectins, "with appropriate stimulation, are decisive for growth response, since they have a reciprocal interaction with the exposed carbohydrate structures of the tumor cells." On the other hand, tumor cells can prevent lymphocytes from identifying them by using hyaluronic acid as a screen (McBride and Bard 1979).

7. Basement Membranes

Basement membranes are a special form of extracellular matrix (Fig. 11). They result from a combination of epithelial cells groups and underlying connective tissue. An underlying basement membrane is imperative for the normal, regular growth of epithelium (Toole 1991). Basement membranes also form a covering over certain types of cells: Schwann cells, terminal "naked" axons, striated muscle fibers, cardiac muscle cells, and smooth muscle cells. Since all of these examples are of types of cells whose membranes depolarize after appropriate stimuli (this opens the Na+, K+, and Ca 2+ channels), the basement membrane with its sugar polymers may represent an important mobile calcium reservoir for these cells.

The basement membrane, which is found under the vascular endothelium, is particularly important as a molecular sieve. This membrane can merge with epithelial basement membranes. There are common basement membrane sections between alveolar epithelium and alveolar capillaries, in the renal glomeruli between the internal leaf of Bowman's capsule (podocytes) and the glomerular capillaries, and as the blood-brain barrier between the capillaries of the CNS and astrocyte processes which are locked together with the perivascular glial lamina. Every alteration of these basement membrane sections results in severe organ damage (e.g., adult respiratory distress syndrome, renal autoimmune disease, oxygen and glucose deficiency in the brain).

Heine (1987b) produced findings as to why the appearance of inflammation does not transfer from the connective tissue to the epithelium. For instance, he demonstrated that in the skin and kidney, the greater vitamin C content of the basement membranes of the epidermis and the renal glomeruli was apparently able to trap the free radicals of an inflammatory process. In Heine's same paper it was shown that in mammary carcinoma, as tumor cells approach the epidermal basement membrane, they lose their vitamin C.

8. Proteoglycans (PGs)

In all organisms, the water-sugar polymer system achieves energy stabilization by attaching to a protein backbone (core) which itself is bound to hyaluronic acid via binding proteins (Figs. 3, 6). Sulfation and amination as well as the addition of acetylneuraminic acid (sialic acid) to the end of the molecule increases the electronegativity of these proteoglycan biopolymers (review Heine and Schaeg 1979).

In order to understand the functional relationships between water molecules and proteoglycan molecules, it is important to become acquainted with their construction (Fig. 6): A proteoglycan molecule is shaped like a bottle brush, the "handle" being a protein core (or backbone) approximately 300 nm long. The "bristles,"—polysaccharide chains approximately 60 nm to 100 nm long, are stretched out because their negative charges repel each other. The molecular weight of proteoglycans is between 106 and 109 Daltons (Hascall and Hascall 1991). Because of this bristle-like configuration, a single proteoglycan molecule can, in terms of molecular weight, eventually take up a vast space (domain), obviously making this molecular configuration especially suitable for binding water. To a large extent, the domain determines the molecular sieve character as well as the visco-elastic, shock absorbing, and energy consuming relationships of the extracellular matrix (summaries: Balasz 1970, Heine and Schaeg 1979, Hay 1991).

The molecular form of tissue water was thoroughly researched by Trincher (1978). The eminent suitability of a network of water molecules for storing information and transferring it between cells, according to Trincher (1978), comes from their molecular structure, which is made of approximately 50% fluid crystals at body temperature. In order to keep water in this state, its minimal energy requirement should be at 37.5°C (Trincher 1978). Misinformation stored in the fluid crystals could be dissolved by appropriately raising the temperature, resulting in a change to a more homogenous fluid (Trincher 1978). Even the bio-photon emission of cells, specified by Popp (1976), could transmit far-reaching informative reactions using these fluid crystal intercellular "bridges."

The formation of parallel molecule clusters arranged in two dimensions distinguishes crystalline fluids, which are limited to small regions and are unstable over time. They are continually being formed and dissolved and, relative to one another, show static, unstructured layers. The size of these clusters is at the scale of light waves. Even weak outer forces are enough to call up a higher state of organization (Hollemann-Richter 1963).

Histological freeze drying shows that the tissue water is in a specific state. During tissue drying, the formation of ice crystals can only be prevented if, when the specimen is rapidly cooled in liquid air (-150°C), the tissue water is not already in ice crystal form. The subsequent drying process must be carried out at a low negative pressure, and below the temperature of the eutectic point of tissue water.

In this way the tissue water changes directly into the water vapor and can be pumped out. If the drying process is stopped too soon, then as the temperature rises there is a gradual breaking of the vacuum that occurs in the temperature range of the eutectic point[2] of the tissue water. (Between approximately -58°C and -52°C there is a short-lasting breakdown of the vacuum with the formation of ice crystals and a destruction of the cell structures. (Heine 1974) The composition by itself does not determine the low eutectic point of tissue water, since media with similar composition, such as is used in cell cultures, begins to freeze when it gets down to -8°C (personal observation). The remarkable eutectic behavior of tissue water appears to depend on the configuration of fluid-crystal water between the sugar polymers of the proteoglycan brushes.

In a multicellular organism, the basis of all intercellular reactions, whether local or far-reaching, is obviously characterized by the water sugar biopolymers of the extracellular matrix, which due to their chemical structure, are capable of conducting and storing information. The system is energetically open and and is capable of disposing of the energy released during all metabolic processes from radical reactions. When passing a certain threshold level, the energy fluctuations derived from these reactions can spread across the extracellular matrix by means of changes in the state of fluid crystal water. These fluctuations can then be used as information by the cells.

The least amount of energy is sufficient for this [reaction to spread], as shown by Pischinger's puncture phenomenon (*Stichphänomenon;* 1983) and Hunke's second phenomenon of neural therapy (1983). The energy displacements initiated in this way do not need to be demonstrable biologically; among other biophysical phenomena, they can be measured as fluctuations of the redox potential of the connective tissue. So in chronic diseases and tumors, which are obviously accompanied by changes in the proteoglycan pattern in the extracellular matrix (Heine 1987), it is logical for the extracellular matrix to show an altered, therapeutic redox potential that the organism cannot usually regulate (Pischinger 1983, Perger 1983).

Consequently the extracellular matrix displays a labile organization system, the main components being sugar biopolymers and water, along with substances dissolved in the water. Here the word, "system," referring to what is apparent, is used to mean "ability," "being-connected-with," or "having-a-mutual-relationship-with"(Gutmann and Resch 1988).

With this definition of "system," what is important is the "system organization" between water and the substances dissolved in it, namely molecules (or ions) of dissolved solids, fluids, and dissolved gasses. In an aqueous solution, hydrophilic substances such as dissolved ions

---

2. A mixture of the constituent substances which produces solidification at a single temperature like a pure substance.

and generally hydrated molecules, e.g., sugar, urea, silicic acid, etc., can be called "structure breakers" (Gutman and Resch 1988). In their surroundings, the sugar biopolymers of the extracellular matrix indeed consolidate the clusters of water molecules into an individual "domain" (Fig. 6). However, the organization of these domains, with regards to any changes, is considerably more apparent in vivo than it would be if the sugar-water system were in pure water.

Conversely, gases dissolved in water, e.g., O2, N2, CO2, or other hydrophobic substances, can be called "structure makers (Gutman and Resch 1988) in reference to the adsorbed water. These authors point out that structure makers" bring out a certain type of arranged dynamism to water structure. Radiological spectra show that gas molecules are embedded in hollow spaces which are larger than than needed to be to accommodate them. This allows the gas molecules a certain freedom of movement. On the inner surfaces of the hollow spaces, for example, the separation between the oxygen molecules is smaller. This generates inner surface tensions which maintain the hollow spaces. Gutman and Resch (1988) further point out that the limited torsional vibration of the gas molecules are able to be tuned to a particular vibratory pattern, and must be rhythmically in unison with the vibratory behavior of the fluid. This again is dependent on the binding behavior of the structure breakers. For instance, the sugar biopolymers of the extracellular matrix. Basically, the reciprocal relationships between structure breakers and structure makers, engage the entire extracellular matrix, even if in different ways and in different body regions. Structure breakers and structure makers thus exercise different, mutually complementary functions in terms of maintaining homeostasis. "In structural imprinting, the structure breakers have a higher hierarchical importance, but in preserving the structural information, the structure makers are the more important." (Gutman and Resch 1988).

8.1. Synthesis of Proteoglycans

The synthesis of parts of proteins and polysaccharides takes place simultaneously at the membrane of the endoplasmic reticulum of the fibrocytes (Fig. 11) (summary Hascall and Hascall 1991). The synthesis of the protein backbone of the PG corresponds to simple proteins and is linked to different types of messenger RNA. Next, a precursor of the protein core is formed in the ribosomes and and passed on to the rough endoplasmic reticulum (ER) (Fig. 11). While the protein core is becoming elongated, the action of glycosyl-transferases in the membranes begins to attach the oligosaccharide chains with N and O bonds. This attachment begins with transfer to the smooth ER cisterns. Gonadotropic hormones obviously take part in this process. The entire synthesis process can be inhibited by tunicamycin (Takatsuki and Tamura 1977).

While the mannose-rich oligosaccharides start their chain growth at asparagine radicals of the protein core through a nitrogen-glycosylamine bond, keratin sulfate chains begin to be bonded to serine (or threonine) through glycosidic bonding of N-acetyl-glactosamine. In the former case, dolichol pyrophosphate intermediates are the initiating, energy-rich bonds, and in the latter, the initial bonding energy comes from uridine diphosphate (UDP, bound to n-acetyl glactosamine.)

The formation of chrondroitin sulfate chains has been researched to a relatively exact degree. The reaction starts with the transfer of the monosaccharide xylose to serine in the protein core through a glycosidic bond. In this process, energy-rich monosaccharides (UDP) are attached to the non-reduced ends of the chains. At the same time, 3'-phosphoadenosine 5'-phosphosulfate (PAPS) forms the biological sulfate carrier. The sulfotransferase activity also takes place in the membrane tubes of the smooth endoplasmic reticulum. Finally, in the golgi apparatus, the PGs are packed into vesicles and extruded with tropocollagen and tropoelastin. The process takes 1–2 minutes from the initial formation of oligosaccharide chains to the release of PGs from the cell into the extracellular space (Dorfmann 1983, Iozzo 1985). Using proline as a radioactive marker, Haus et al. (1968) observed a new synthesis of collagen within minutes after noise stress. With a radioactively marked sulfate as a precursor in an experimental stress model, Heine and Henrich (1980) showed that extracellular matrix, synthesized from myocytes from the tunica media of a rat's femoral artery, shows a significant increase in sulfated extracellular matrix components (chondroitin sulfate and dermatin sulfate proteins) after one hour of sympathetic stimulation. Muscle cells of the vascular wall are derived from fibrocytes, and are also capable of extracellular matrix synthesis. Dermatin sulphate in particular is able to trap lipids and calcium ions, which in turn encourages arteriosclerotic changes.

If the stress is continuous, stress-induced synthesis of ground substance must eventually lead to functional disturbances in the extra cellular matrix and hence to changes in its molecular sieve character. These events finally lead to a vicious cycle, laying the groundwork for chronic illnesses. After experimental multiple traumas in dogs, and also on examining lung biopsies of severely traumatized accident victims, Heine et al. (1980) showed that there is an increase in collagen in in the alveolar septa within 30 minutes, and weeks later this leads to the development of adult respiratory distress.

The unique structural properties of PGs can be seen in the example of a monomer of a chrondroitin sulfate proteoglycan molecule (CSPG), in which the protein backbone exhibits four distinct regions (Iozzo 1985):

1. A binding region for hyaluronic acid. It consists of two binding proteins (a smaller one with a molecular weight of 42,000 D, and a larger one with 50,000 D; Baker and Caterson 1979). These 'link proteins' do not have any sugar side chains.
2. The region of the N terminal is attached to the above, with 10 to 15 N-glycoside-bound oligosaccharide chains.
3. The keratin sulfate binding region follows distally. It takes up approximately 10% of the protein core.
4. Following this is the chondroitin sulfate binding region, which extends

over approximately 60% of the protein core.

Despite its relative size, the mass of the protein core is only 5–10% of the total mass of a single PG monomer. Approximately 100 chondroitin sulfate chains and 50 to 60 keratin sulfate chains have O-glycoside bonds to the protein core of a monomer. Distributed over the rest of the protein core are another 50 to 60 O-bound oligosaccharides of the mucin type, i.e., they have terminal sialic acid groups.

In the articular cartilage of humans and in the nasal cartilage of cattle, the chondroitin sulfate binding region contains bound covalent phosphate groups, in the ratio of one phosphate chain to 20 oligosaccharide chains. The preferred site for phosphorylation appears to be the C2 atom of xylose. In swarm chondrosarcoma in rats, about 80% of the chondroitin sulfate chains are phosphorylated (summary Iozzo 1985).

A CSPG molecule with a length of approximately 300 nm has a molecular weight of about 250,000 D. The oligosaccharide side chains are extended and approximately 50–60 nm in length. Each single chain has a molecular weight of about 20,000 D. The distance between the oligosaccharide chains is about 30 nm., obviously due to the (negative) charge. The total molecular weight of a monomer PG molecule is approximately 2.5×10^6 (Hascall and Hascall 1991). With great specificity, monomer PGs attach to a long (up to 1 mm) hyaluronic acid group (Fig. 6) by means of their binding proteins. In doing so, the spacing of about 30 nm is maintained. This spacing corresponds to a disaccharide unit of hyaluronic acid which repeats itself after five other disaccharide units. The disassociation constant is 10^{-6} to 10^{-7} M, which is extremely stable (Christner et al., 1979). Up to now there is no known anionic biopolymer that would compete for these binding sites. An aggregate of PG and hyaluronic acid contains about 40 PG monomers and has a molecular weight of approximately 10^8 D (Hascall and Hascall 1991). The half-life of hyaluronic acid is 6 to 12 days, that of chondroitin sulfate 7–17 days, and that of keratin sulfate protein 60–120 days. Growth hormones and estrogens elongate chondroitin sulfate protein chains and shorten keratin sulfate protein chains.

8.2. Functional Aspects of CSPG-Hyaluronic Acid Complexes

As described, CSPGs (chondroitin sulfate proteoglycans) together with hyaluronic acid form large supermolecular complexes, up to 1 mm long. Relative to their molecular weight they occupy a huge hydrodynamic volume (domain) (Fig. 6). If the volume of the tissue water is decreased, there is a stronger intermolecular interaction among the stiff polyanionic oligosaccharide chains. When water is added, they again extend.

Since the PGs and hyaluronic acid give the extracellular matrix a net-like supermolecular structure, cover cell surfaces, form the intercellular substance, and ensheath and permeate elastic and collagen fibers, this buffering, viscoelastic property of PGs and hyaluronic acid is extremely important for normal tissue and cell function. This particularly helps in understanding how joint cartilages function and how heart and blood vessels pulsate. The local PG and hyaluronic acid concentration in a tissue region will therefore determine its mechanical dynamics.

One must always keep the following facts in mind: endogenous errors or exogenous factors can alter or disturb the single steps in synthesis and metabolism of PG as well as its protein core, its oligosaccharide side chains, its binding to hyaluronic acid, the free GAGs, and also the structural and network proteins (collagen, elastin, fibronectin, etc.). Endogenous errors include familial or aging-associated genetic errors or mistakes in the cellular control mechanisms. Exogenous factors include environmental burdens—especially free radicals, malnutrition, stress, viral and bacterial infections, radiation, alkalizing substances, and so on. Both endogenous errors and exogenous factors can cause a tremendous number and variety of connective tissue, vascular, cartilage, and bone diseases. The equilibrium of the biological flow of PGs is mainly determined through proteolytic enzymes, especially the serine proteases such as plasmin, since most of the oligosaccharide chains attach to the protein core of a PG via serine. Hydrolytic enzymes split the GAGs and the oligosaccharide chains. The lysosomes of neutrophil granulocytes and macrophages contain the greatest variety of these enzymes. This casts a special light on normal leukocytolysis as an important regulator of both formation and breakdown of the extracellular matrix.

8.2.1. Dermatin Sulfate Proteoglycan

In contrast to CSPG, dermatin sulfate proteoglycan (DSPG) is not bound to hyaluronic acid. The protein core is very long and has a molecular weight of $\leq 300,000$ D). In contrast to CSPG, it has about 10 to 20 oligosaccharide side chains, which all contain iduronic acid. The molecular weight of the side chains of a DSPG molecule is more than double that of the CSPG side chains (about 56,000 D). DSPG also contains more neuraminic acid acid. The protein core has 300 to 400 oligosaccharide side chains of the mucin type with O-glycosidic bonds, and about 50 side chains with N-glycosidic bonds, which are attached to serine and threonine at the core. Thus DSPG is very similar to the mucus produced by the glandular epithelia ('mucins') (Summary Iozzo 1985).

8.2.2. Functional Aspects of Dermatin Sulfate Proteoglycan

In contrast to the more densely packed CSPG which is attached to hyaluronic acid, DSPG has an more open structure which develops into even larger domains. DSPG appears in larger amounts in the extracellular matrix of the connective tissue of the skin, cornea, sclera, tendons, cartilage, and bones. DSPG has a special function in follicle ripening and ovulation. It becomes increasingly concentrated in the fluid of the cavities in the tertiary follicle (Graafian follicle), keeps this under tension, and gives the follicle liquid the necessary viscosity for ovulation. If gonadotropic hormones cause the amount of follicle liquid to increase, the follicle epithelial cells synthesize more DSPG (Iozzo 1985).

DSPG plays a significant role in the maturing of collagen fibrils. It binds in the area of the D-band (gap region of the collagen fibril) and through this it couples the fibrils to one another into a higher order of fibers (Öbrink, 1975).

8.3. Heparan Sulfate Proteoglycan

Like heparan sulfate as a GAG, heparan sulfate proteoglycan molecules (HSPG) have a special affinity to cell surfaces. More exact structural studies were carried out on the HSPG of human colon carcinoma cells (Iozzo 1985). The protein core has a molecular weight of approximately 240,000 D. Besides about 100 to 200 oligosaccharide chains with O-glycosidic bonds to the protein core, which contain glucosamine, glactosamine, and sialic acid, there are 10 to 15 pure heparan sulfate chains. In contrast to all other PG types, the protein core of HSPG contains a hydrophobic domain that anchors HSPG in the cell membrane. The bond is very strong and can only be dissolved through proteolysis. It is possible that this anchoring location extends into the cytoplasm of the cell and interacts there with the filaments of the cytoskeleton (Iozzo 1985).

8.3.1. Functional Aspects

Besides this important transmembranous aspect, HSPG has further potentials for attaching: The heparan sulfate side chains of the PGs can bind to 'receptors' in the glycocalyx by forming ions. These bonds can be dissolved by adding polyanions such as heparin. In this manner, in vitro, it was calculated that there are 6×10^7 endothelial bonding sites per cm2 for heparin (Psuya et al. 1987). Heparan sulfate chains of a variety of monomers can bond with one another. Since all PG types have one or more heparan sulfate side chains, HSPG, and heparan sulfate have an important function in the maintenance of cell groups and their anchoring to the basement membranes, that is, to their structural components (collagen, fibronectin, laminin). It is assumed that these polymers directly influence cell growth and multiplication because of these "surface-associated" properties of HSPG and heparan sulphate (Chiarugi 35 al. 1976, Iozzo 1985).

8.4. Keratan Sulfate Proteoglycan

What is special, structurally, about keratan sulfate proteoglycan (KSPS) is that its oligosaccharide side chains contain D-galactose, and that none of the side chains are bound via xylose to the amino acid serine of the protein core. In fact, there are O-glycoside bonds between N-acetylgalactosamine and the hydroxyl groups of serine or threonine. In the cornea there are N-glycoside bonds between N-acetylglucosamine and asparagine.

8.4.1. Functional Aspects

Little is known about the physiological role of KSPG. In the second half of life there is an increase in KSPS as well as collagen and chondroitin 6–sulfate PG. Patients with macular corneal dystrophy cannot form any mature KSPG (Hascall et al. 1980).

The PGs are no longer designated by the type of their GAG-chains, but according to the tissue localization and their interim protein core composition (the sequence of amino acids without considering the GAG chains) (Tab. 9; summary by Heine 1997).

9. Structural Glycoproteins

9.1. Collagen Synthesis, Molecular and Supramolecular Structure

With the exception of blood cells, all cell types—at least up to a certain level of development—seem to be capable of collagen synthesis (Hay summary 1980).

It is difficult to ignore the genetic control of collagen synthesis. In addition to those genes that determine the primary structure, other genes must simultaneously provide a code for enzymes. These enzymes, for instance, control the correct length of a triple helix molecule, its glycosylation, crosslinking, and polymerization in the cell, as well as the exit of the procollagen molecule (tropocollagen) into the extracellular space. GAGs and PGs are significantly involved in the extracellular maturation of collagen fibrils and fibers (Fig. 11).

The intracellular biosynthesis of collagen includes the following steps: the amino acid sequence for three single α 1 chains (primary structure) of triple-helix molecules is determined in the various collagen-producing cells through the formation of a complementary messenger RNA, which is transcribed at the appropriate DNA segments of certain chromosomes (human chromosome 7 and 17 for type 1 procollagen of the cornea and type 1 procollagen of human skin; summary by Hay 1980). (400,000 base pairs of DNA were investigated for the pro-α_2 gene, which contains approximately 50 exons [coded regions], that are interrupted by introns [non-coded sequences]). Procollagen molecules that correspond to the DNA coding, are assembled from pro-α1 chains at the ribosomes of the endoplasmic reticulum (ER) when transfer-RNA molecules (each of which introduces the appropriate amino acid) are temporarily added. It appears that the 'signal peptides' at the end of the procollagen molecules enable the molecules to penetrate from the ribosomes into the tubule system of the endoplasmic reticulum (pre-procollagen'). The signal peptides are removed as they cross the membrane (of the ER). In the cell, the α1 chains are each formed into a helix as far as the Golgi apparatus, which is the terminal and packing station of the endoplasmic reticulum. Each final helix formation is made up of of three α1 helical chains which then interlock into a triple helix (tertiary structure; Fig. 11). Except at the COOH and NH_2 ends which are non-helical, the three α1 chains are again wound around one another and form a superhelix.

Hydrogen bonds, which connect the internal structure of the individual α1 chains and form bridges between them, are involved in the development of the secondary and tertiary structures. Glycine plays a special role in the formation and stabilization of these bonds. Every third amino acid along the α-polypeptide chain is a glycine. Hydrolysine has a particular influence on the stability of triple helix formation, replacing approximately every hundredth proline as an enzymatically-hydroxylated proline (Hay). It is important to note that only during the formation of the α1 chain are 1/3 of the proline and 1/6 of the lysine 1/3 hydroxylated, through the enzyme system (cofactors involved in this: oxygen, ascorbic acid, α1 ketoglutarate, and iron (Fe^{2+} ions). At the same time some of he hydroxyl groups of the hydroxylysine are attached to the galactosyl-glucose or galactose groups (allowing hydrogen bonds to form between the triple helices). Collagen is thus a glycoprotein, and functionally it is a structural glycoprotein. The high proportion of amino acids glycine (23–29%), proline (15–16%), and hydroxyproline (11–14%) is striking. The helix twists uniformly, because glycine, the smallest amino acid, and other amino acids repeat themselves in groups of three (approximately 333 glycine fragments per chain). The sequence of amino acids in the α1 chains is thus Gly-X-Y, where X or Y can be any other amino acid. It is most common (in approximately 100 triplets on an α1 chain) for X to be a proline and Y a hydroxyproline (Hay 1980). Given a total of 1000 amino acid fragments, 81 are negatively and 82 are positively charged, so collagen has an almost neutral electrical charge (for piezoelectricity, see the text below as well as the contribution by Bergsmann).

The formation of an α1 chain takes about ten minutes. The complete synthesis of a molecule, including leaving the cell, takes 35 to 40 minutes (Olsen 1983). The triple helix collagen molecule has an extracellular length of approximately 280 nm (2800 Å) and a diameter of 1.5 nm. The triple helix structure is stable at 37°, but above 40°C it begins to unwind (Harkness 1970). At the NH- and COOH-ends, the telopeptides, which were never wound into the helix shape, are partly split off by a procollagen protease while the triple helix is being released into the extracellular space. [Before leaving the cell] the telopeptides are bound to each other with disulfide bridges and are crosslinked by glycolized hydroxylysine. Thus, while inside the cell, the telopeptides help stabilize the procollagen molecules and prevent fibrils from forming prematurely. During release into the extracellular space, the telopeptides that split off act as a kind of negative feedback of inhibitors to collagen synthesis. Since the terminal crosslinks appear on the outside of the triple helix configuration, these crosslinks are much more accessible to proteolysis than those inside it. Lastly, only collagenases from macrophages and neutrophilic granulocytes can attack the extremely stable collagen molecule (Hay 1980).

[The triple helix molecules are arranged in rows. In each row there is a space between the tail of one molecule and the head of the next.] The rows are parallel to each other, and the ends of the molecules overlap each other [in a staggered pattern] by about 27 nm, i.e., about one quarter of the length of each molecule (Fig. 11). This leaves a hollow gap of about 40 nm in diameter between the staggered molecules, that is, between the NH_2 head and the COOH tail. Not only do the glycosyl derivatives of this hollow area take part in the crosslinking of the collagen molecule, but they also are able, for instance, to bind water and exchange cations. This means that the hollow sections become the "start region" of the ossification processes, and are available for the storage of hydroxyapatite crystals (summary by Hay 1980, Linsenmeyer 1991).

9.1.1. Collagen Modification

Since hydroxylysine and hydroxyproline are present in collagen molecules in varying amounts and they can be glycolized differently, the result is a great variety of modifications of the collagen molecule. The various types of messenger RNA as well as various types of disulfide bridges of the telopeptides can also lead to specific differences in molecular construction. Presently, twelve well-characterized collagen molecule types have been differentiated (Linsenmeyer 1991); only the five most important ones are presented here.

1. Type I collagen is the most widely distributed in connective tissue, including in bone and dentin. Each molecule has two α1 chains and one α2 chain. The formula is [α1 (I)] - α2 (I).
2. Type II collagen mainly occurs in cartilage. It consists of three identical α1 chains, and so its formula is [α1 (II)]3. Both Type I and type II collagen can polymerize in the same fibrils, as was shown in the cornea of chickens (co-polymerization).
3. Type III collagen is an important structural glycoprotein, which interlaces the extracellular matrix as a fine "reticular network. It is comprised of three α1 chains. Its formula is [α1 (III)]3. Type III collagen contains a great deal of 4–hydroxyproline, glycine, as well as 2 cystine remnants per α chain.
4. Type-IV collagen is a part of basement membranes. There are nonhelical regions in the molecule (Kefaldis 1978).
5. Type V collagen is also limited to basement membranes. It is especially easy for a collagenase from tumor cells to attack this type of collagen.

Until now, awareness of these collagen types has significantly clarified connective tissue diseases (e.g., Marfan's Syndrome, osteogenesis imperfecta, chondrodystrophy).

In positive contrast specimens stained with phosphotungstic acid, typical striations with 12 bands appear during the side-to-side polymerization of tropocollagen molecules to collagen fibrils, from about 10 nm in diameter. The bands correspond to polar segments, and the space between to the apolar regions of a microfibril. This banded arrangement is connected to the displacement [of successive rows of molecules] by approximately 1/4 of the length of each molecule in the primary fibril. The length of each repeated unit is approximately 67 nm (Fig. 11). The biological half-life varies between 1 and 630 days (Olsen 1991). The fibrils (diameter 10–25 nm) aggregate into fibers (diameter about 0.3 μ), and these then aggregate to strong fiber bundles several millimeters thick (Fig. 11). In the extracellular matrix, there are always broken-down tropocollagen molecules that can contribute to growth in the thickness of the fiber bundles. This molecular breakdown can be precipitated by amino acids (e.g., valine from bacteria). Under pathological conditions, long-spacing collagen molecules can also appear which are lined up end-to-end and head-to-head (with no staggering between the rows) with a periodicity of up to 280nm or globular configurations (Fig. 11).

9.1.2. Functional Aspects

The thickness of the collagen fibrils is organ-specific and changes with age. Non-enzymatic glycosylation with a tendency towards ionic and lipoprotein bonding plays a major role in the aging process of collagen.

In the tendons, collagen fibrils are arranged lengthwise. Because of this, they have only one-tenth the tensile strength of skin, which has fiber bundles going in all directions. The quality of the fibers plays a large part in this. Tendons are made of Type I collagen, whereas all types of collagen are found in skin. Besides this, the fiber system in the skin first becomes aligned in the direction of tension before the individual fibers are loaded.

The bonding of sugar to the tropocollagen molecule [especially in the area of lacunae. See above.] and the association of proteoglycans of the ground substance with the surface of fibrils and fibers (Henle's sheath) create a two-phase system which is capable of binding water, and that, in turn, stabilizes and orients the fibers. Through low elasticity, relative resistance to tearing, and a highly asymmetric length-to-width relationship, collagen adds significantly to the viscosity of the extracellular matrix, apart from the GAGs and PGs. Globular molecules demonstrate this effect on a disproportionately smaller scale. This effect is easy to recognize when a collagen solution, still in its naturally asymmetric state within the molecule, is heated above the denaturing temperature of collagen (>40ºC). The viscosity of the solution then decreases rapidly, and only reappears when it is again cooled below this level.

The *antigenicity* of the collagen types is relatively low, with tyrosine possibly playing a leading role (Schmitt et al. 1964, Linsenmeyer 1991). Procollagen shows a higher antigenicity than tropocollagen (Summary in Olsen 1991). A very important, but until now little-noticed quality of collagen is the piezoelectric property (Athenstaedt 1969); article by Bergsmann, p. 91) The piezoelectric property here refers to the electrical charge that results from deformation under a mechanical load (pressure, tension, torsion).

As Athenstaedt (1969) showed, piezoelectric forces already play a significant role in the orientation and polymerization into fibrils of the tropocollagen molecules when they are extruded from the fibrocytes. However, not only does a permanent electrical polarization appear in a longitudinal direction, but also at right angles. It is easy to explain the electrical polarization in a longitudinal direction. This depends on the asymmetrical morphological structure of the tropocollagen molecule. The polar and apolar amino acids are distributed asymmetrically along the molecular axis. Some carry an excess of positive charges, others an excess of negative charges. The charge center does not match the molecular axis, but lies a certain distance from it. As a result, the tropocollagen molecule behaves like a dipole, with one direction along the molecular axis and the other at right angles to it. The single-axis optical double refraction of the collagen fibril corresponds to this. On the other hand, it is difficult to explain the permanent electrical polarization at a right angle to the direction of the fibrils. According to Athenstaedt (1969) this is not coupled to the influence of neighboring tissue or to the direction of the collagen fibers that sheer away from the general direction of the fibers. In addition, the hexagonal packing of collagen fibrils inside a fiber is too strong for a piezoelectric effect to result. It was also thought that the proteoglycans embedded in the fibrils could cause the effect. However, the same piezoelectric effect is present in bones dating from the ice age, even though the proteoglycans have long since been destroyed. The electric polarization of collagen at right angles to the orientation of the fibers must therefore rest with the ultrastructure of the collagen molecule itself. In the apolar region of a microfibril, a hexagonal packing of the tropocollagen molecule appears. In the polar region, with the amino acids glutamin, asparagine, lysine, and arginine, there is no hexagonal packing, and correspondingly there is no marked electric polarization at these sites. The crosslinking of tropocollagen molecules is, however, brought about to a great extent through the chemically active side chains of the polar amino acids. The electric polarization at right angles to the longitudinal fiber axis could be tied in with these side chains, if one assumes—and there are findings available on this (radiological structural analysis, double refraction of swollen collagen fibers)—that the side chains are so oriented along the axis of the fibrils that no rotational symmetry appears, but instead of that there is an excess of positive charges on one side and an excess of negative charges on the other. Concurrently, this means that the strength and number of the cross-links on one side of the microfibrils must be greater than on the other (Athenstaedt 1969)! An important consequence of this is that, for instance during the healing of a fracture, the cells of callus tissue and the adjacent bone obviously "understand" this electrical potential pattern across their cell membranes and regulate the structural remodeling of the support structures accordingly (summary Heine 1997).

9.2. Elastin Synthesis, Molecular and Supramolecular Structure

Elastin synthesis follows the same principles as those described for collagen. Here too, fibroblasts and smooth muscle cells of the vascular wall are capable of elastin synthesis. As in collagen synthesis, a soluble proelastin (tropoelastin) is released into the extracellular space, and aggregates into insoluble elastic fibrils through cross-linking. These fibrils mature further into branched fibers, which only have a sugar coating on their surfaces (Heine and Schaeg 1979, Fig. 1). Elastin itself does not contain any neutral sugars, hexosamines, cystine, or tryptophan (Franzblau and Faris 1983).

In the same way that collagen molecules have repeating Gly-X-Y amino acid triplets, proelastin has a Gly-Val-Pro-Gly tetrapeptid. Changes in the sequence result in elastin molecules which are correspondingly different. And indeed the high glycine and hydroxyproline content could be an indication that elastin have developed from collagen (Heine and Schaeg 1979).

During the extracellular maturation processes of tropoelastin, which in contrast to collagen maturation tends to be inhibited by ascorbic acid, there is development of desmosine and isodesmosine, the typical amino acids of elastin. They are cross-linked products of lysine. The deamination of the lysine amino group through lysil oxidase initially leads to lysine aldehyde (allysine). (All agents that participate in this step hinder the production of the cross-links. Two of these agents are D-penicillamin or the lathryogen β-aminoproprionitril.) With aldehyde remnants of the same or adjacent tropoelastin molecules, allysine can undergo an aldol condensation, and this results in desmosine. Alternatively, via its amino group, non-oxidized lysine together with allysine can form a Schiff base, resulting in isodesmosine. Desmosine and isodesmosine form the sugar-free core of elastic fibers. Around the amorphous center of elastic fibers, there are, however, glycosylated microfibrils (diameter about 8 nm). In contrast to the amorphous elastin, these microfibrils have a high cystine content (70–80 cystine remnants per 1,000 amino acids). The resultant cystine-disulfide links are responsible for the relative insolubility of the microfibrils. According to Scherr et al. (1973), the amino acid sequence of the microfibrils corresponds remarkably with the terminal NH- and COOH-chains of the procollagen molecules. The microfibrils guide the orientation and the form of the maturing fibers, and thus also the formation of the cross-links. Hydroxyproline, hydroxylysine, and lysine do not take part in the cross-links (Franzblau and Faris 1991). Special, immature, very fine elastic fibrils (diameter about 8–10 nm) are called oxytalan fibers. They help anchor the basement membrane in the ground substance. Along with Type III collagen, oxytalan fibers make up the optically visible part of the basement membrane. This mat-like meshwork of Type III collagen and oxytalan fibers also sheathes the parencymal cells as a sort of fiber stocking, which on a microscopic level, keeps them in "shape."

9.2.1. Functional Aspects

Elastic fibers are the foundation of the flexibility and the normal tone of the skin, the lungs, and the blood vessels. Like collagen, elastin has its greatest functional potential, its rubberlike flexibility, at 37°C. According to experiments, at 20°C, elastin takes on a peculiar glass-like consistency, becomes brittle, and breaks (summary by Franzblau and Faris 1991). According to Hoeve (1977), lipid deposits in the blood vessel walls bring the "glass point" close to 37°C. Elastic fibers can be stretched up to 130% of their initial length, which is 20–30 times more than collagen. Elastic fibers form three-dimensional nets, which are not affected by heat, acids, or alkalis. Their number decreases with a person's age, especially in arterial walls afflicted by arteriosclerosis.

Trypsin and pepsin break down elastin more easily than they do collagen. Elastases have been isolated from neutrophil granulocytes, macrophages, thrombocytes, and pancreatic tissue (summary by Olsen 1991). These enzymes, however, are not specific. While cellular elastase has a high proteolytic activity on leukylpeptide bonds, pancreatic elastase specifically acts on adenylpeptide bonds.

Excessive elastolytic activity occurs, for instance, during pulmonary emphysema, pancreatitis, and arteriosclerosis (Franzblau and Faris 1991). The proteolytic enzymes bromelain, papain, ficin, pronase, and nagase also attack elastic fibers (Franzblau 1971).

The relationships between elastin and collagen determine the mechanical properties of connective tissue. Elastic-fiber nets form an irregular coil-like wrapping around collagen fibers, creating the typical ripple that collagen fibers show. In traction-tension diagrams of fresh elastic fibers, there is a point from which the measurements of tensile strength suddenly increase.

The arrangement of the elastic fiber meshworks along the collagen fibers can aid a morphologically understanding of this sudden increase in tensile strength, because the ripples of the collagen fibers serve as a reserve. The viscoelasticity of a tissue can be measured by how long this phenomenon takes. This is simultaneous expression for conformational changes in macromolecules that have a potential for intermolecular shift (summary by Franzblau and Faris 1991).

Irregular-conformation changes have been described for mucopolysaccharides. For example, as opposed to normal relationships in the aortic wall, in Marfan's syndrome major changes in the tension-traction curve appear in the longitudinal and transverse directions (Yamada 1991). This causes clinical patterns such as aneurysms, aneurysm dessecans, or dilatation of the aortic valve to develop. It is assumed as a matter of principle that all the pathological changes in the extracellular matrix are accompanied by conformation changes of its macromolecules.

From an energy point of view, a state on a higher order means a decrease in entropy (random molecule distribution). A lower order thus implies an increase in entropy, which means a greatly increased possibility that the fibrillar molecules will be deformed in their assembly and in the direction of their orientation.

Gosline (1976) discussed whether the reactions (Coulomb charges) of the hydrophobic groups of elastic fibers with water molecules could store energy during the stretching phase. This energy could be freed again during return to the original form. It can be assumed that through interactions with water molecules, Coulomb charges significantly maintain the collagen-elastin-PG-GAG interconnection, and that they control its viscoelastic behavior.

Elastic fibers in various tissues differ in quality and quantity (review Olsen 1991). The ligamentum nuchae of cattle is 80% elastin. The fibers are very thick (6.7 µm), and consist of short, rod-shaped fibers. In the elastic type of arteries, the elastin content is about 50%. In contrast, in skin, loose connective tissue and tendons, elastin makes up 5% to 2% of the dry weight. In the tunica media of the human aorta about 50–60 elastic membranes are arranged in concentric layers in the tunica media of the human aorta.

They are 2.5 μm thick and separated by connective tissue spaces 6–18 μm wide, and which are filled with extracellular matrix that contains smooth muscle cells and frameworks of collagen and elastic fibers.

9.3. Crosslinked Proteins

At present, fibronectin, laminin, and chondonectin have been differentiated. These macromolecular, sialinized glycoproteins can all mediate between cell surfaces and the extracellular matrix because they are able to bind collagen, and they are all exceptionally easy to split by proteolysis (Summary Yamada 1991, Haynes 1983).

9.3.1 Fibronectin

The fibronectin molecule was discovered on cell surfaces (Haynes 1983), and then found to be a 'tissue adhesive' in the extracellular matrix. It was then finally recognized as an important component of blood plasma (300 μg/ml). It was also demonstrated in cerebrospinal fluid (3 μg/ml) and in amniotic fluid (60–80 μg/ml). Except for erythrocytes and lymphocytes, fibronectin seems to be produced by all types of cells, particularly by fibrocytes, macrophages, granulocytes, and epithelial cells. It appears at a very early stage of embryonic development, and has been found in mice as early as the morula stage (Toole 1983).

As a monomer molecule, tissue fibronectin has a molecular weight of 220,000–240,000 D and a carbohydrate content of 5%. It appears as a monomer, dimer, and polymer molecule (Fig. 12). Fibronectin is found dissolved in plasma as a disc- or rod-shaped dimeric molecule. The fibronectin subunits are highly asymmetrical. No α-helical or β-sheet formation has been observed. The molecule is constructed from polypeptide domains, and is bound by the folded polypeptide chains. They can stretch or shorten according to the relationship of ions in the environment. Each monomer molecule has five sugars (n-acetylglucosamine, mannose, galactose, fucose, sialinic acid) bound to asparagine. The collagen- and cell-binding part of the molecule is characterized by oligosaccharide chains. The half-life of fibronectin, like that of protein, is generally 30–36 hours. The fibronectin molecule has various molecular-specific binding sites (Fig. 12). Close to the carboxyl end is a collagen-binding fragment. Between this and the amino (NH-) end of the molecule is the fragment that binds to the cell surfaces. This fragment includes binding sites for heparan sulfate, heparin, hyaluronic acid, and proteoglycans. Between the binding sites for heparin (heparan sulfate) and collagen, there is an actin binding area. This is especially important because according to findings, fibronectin can have a direct relationship across cell membranes with the actin of the microfilament bundle of the cytoskeleton. Cytochalasin A and B disrupting the microfilament system in the cytoplasm leads to a dissolution of the fibronectins bound to the cell surfaces (Yamada 1991).

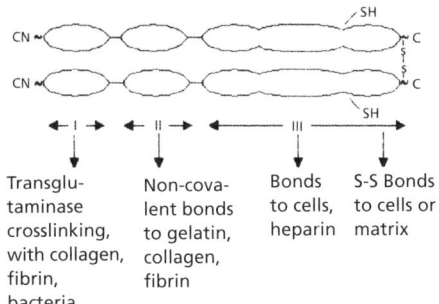

Figure 12. Diagram of the structure of fibronectin. The regions marked by dashes can be split by pepsin. The arms are about 2 nm in diameter and have larger globular components at their ends (about. 5–7 nm) (after Timpl and Martin, from Yamada 1991).

9.3.1.1. Functional Aspects

Fibronectin takes part in all growth, differentiation, and movement processes of the cell. It mediates in the adhesion of cells to the basement membranes, also in intercellular contact, which prevents a reciprocal migration of cells. (Tumor cells lack this contact inhibition.) Fibronectin enmeshes the macromolecules of the extracellular matrix with each other and with the glycocalyx of cell membranes. Both inside and outside of blood vessels, fibronectin can be covalently crosslinked through the enzyme transglutaminase (Factor XIII of blood coagulation). This can cause fibronectin to form complexes both with itself and with collagen and fibrils. Heparin together with fibronectin can also form microfibrils (summary by Hay 1980).

Since at its actin binding area, fibronectin can bond across cell membranes to the cytoplasmic microfilament system ("cytoskeleton"), this allows multifaceted extracellular-to-intracellular information paths to materialize (as well as intra- to-extracellular paths), which enable cells and cell groups to react rapidly to changes in homeostasis. In contrast, tumor cells show an irregular cytoskeleton. Accordingly, certain peculiar qualities of the glycocalyx formation and fibronectin bonds appear which make it more hard for these cells to recognize each other, which in turn leads to a decrease in contact inhibition. Because fibronectin is very sensitive to proteases, these information couplings of cells-to-matrix can be severely disturbed through an abnormal rise in proteases (e.g., all types of inflammations, shock, rheumatism). In any case, the sugar fragment of fibronectin prevents a too-rapid proteolitic splitting and a too-rapid turnover of the polypeptide chains (Yamada 1991, Heine and Domann 1984).

Animal experiments show that within ten minutes after the onset of a traumatic event, fibronectin is already intra- and extravasally diminished through proteolysis, which triggers a general cellular alarm reaction (Heine and Domann 1984). On the other hand, it has been observed in embryonic myotubes that fibronectin is important in the transformation of myotubes to striated muscle fibers. As embryo development proceeds, specks of fibronectin regions remain on the plasma membrane of the muscle fibers. These specks appear to be the starting points for the development of myoneural synapses (motor end-plate) (review by Sengbusch 1979). Supporting this theory is the example of acetylcholinesterase, which breaks down the neurotransmitter acetylcholine in the myoneural synapses. Acetylcholinesterase has a final fragment that is related to collagen. It is therefore conceivable that, through fibronectin, the enzyme can bond to the surface of the muscle fibers in the region of the postsynaptic membrane section and can, accordingly, have an effect in transmitting information; that is, fibronectin can add to the development and renewal of these synapse types. Since proteoglycans and fibronectin appear together in the extracellular matrix, and since certain proteoglycans are responsible for the direction of the sprouting of nerve fibers, fibronectin seems to take part in the sprouting of axons both in the central nervous system as well as in the periphery (summaries by Sengbusch 1979, Toole 1991).

In plasma and in the ground substance, fibronectin also has the function of an "opsonin" for antigens; this means that after fibronectin bonds to the antigen surfaces, it is identified as foreign by macrophages and neutrophil granulocytes, and is ingested, and destroyed by phagocytosis. Since complement factor C1q has a terminal part related to collagen, it also becomes an opsonin after it binds to fibronectin. It is significant that the actin-binding region of fibronectin can also bind DNA. If DNA is released from broken-down cells, in large amounts (which happens in cases of tumors and shock), it can be opsonized through fibronectin and rendered harmless (Yamada 1991).

Fibronectin has special functions in faulty epithelium. First, thrombocytes cover the defect and release a series of substances that lead to the further binding of platelets. This sets off the cascade of events leading to coagulation, and fibrin is precipitated locally, whereupon an insoluble thrombus is formed by enzymatic crosslinking of transglutaminase with fibrin. The fibronectin in this process comes from the plasma, the endothelial basement membrane, the extracellular matrix, and the broken-down endothelial cells. On the side facing the lumen of the blood vessel, endothelial cells contain only a little fibronectin, but a great deal of heparin. Endothelial cells can even pump heparin against a gradient of 100:1 from the extracellular matrix to their surfaces (Jaques 1980). As a result, the surface of the endothelium facing the lumen has a strong negative charge, and this in turn increases the antithrombotic effect (Heine 1986). On the other hand, fibronectin can aggregate heparin into fibrils, thus supporting thrombus formation (Haynes 1991).

9.3.2. Laminin

This glycoprotein only occurs in basement membranes. According to Timpl et al. (1978), laminin consists of two polypeptide chains, each with a molecular weight of 220,000 D. Disulfide bridges bond the two chains to each other. The carbohydrate content is about 20%, and 4–6% of this is sialic acid. The macromolecule consists of one long and three short arms with globular endings (Fig. 13). Fibrocytes do not form laminin.

The main sources of laminin are epithelium and striated muscle fibers. The main function of laminin is to assist in the adhesion of epithelia to the collagen (type IV) of the basement membrane. It also binds to heparin and heparin sulfate, both of which are also constituents of the basement membranes (summary Haynes 1991).

9.3.3. Chondronectin

Hewitt et al. (1980) discovered this glycoprotein (molecular weight about 180,000 D) in cartilage and serum. In mesenchymal tissue that is transforming into cartilage, chondronectin is replaced by fibronectin.

Functionally, chondronectin is an adhesion factor for chondrocytes in cartilagenous collagen (type II) (summary Haynes 1991).

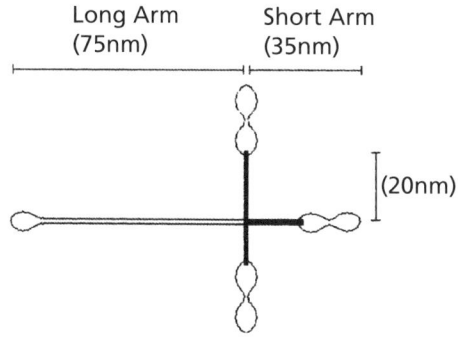

Figure 13. Diagram of the structure of a dimer laminin molecule with its functional domains (after Haynes 1991).

10. Energy Flow in the Extracellular Matrix

A description of the functional structure of the extracellular matrix can clarify the logical basis of certain life forms, but at this time, the prime mover is still no clearer. The common denominator of all biological reactivity is its energy flow. The material basis of biological energy is comprised of the various sugar biopolymers, water, the substances dissolved in the water and the degree to which they are structured. Biological energy can be understood quantitatively and qualitatively as biodynamics (reactions between biological objects) and bioenergetics (transformations in quality between energy and mass (*Qualitätstransformationen*). A very clear example of this is the interaction between water and the extracellular matrix, in which the energy flow between the molecular clusters of fluid-crystal water and sugar molecules determine the degree of organization and thus the degree of structure of the extracellular matrix.

Since all life functions are mediated by the extracellular matrix, this matrix cannot consume more energy than specialized systems of a higher order, such as cells and cell groups. The type of system involved in the case of the extracellular matrix is a dissipative system whose states of organization are unstable. This system fluctuates widely, far from a thermodynamic equilibrium. In such a dissipative system, spontaneous reactions in the extracellular matrix trigger new states of order, and spread by acting as catalysts themselves (autocatalysis). Thus these spontaneous reactions and new states of order are the expression of basic life-maintaining reactions and of transformations in quality between energy and mass. When the free energy of the products is lower than that of the reactants, these processes appear spontaneously.

The free energy exchange of a system (ΔG) is the sum of all kinetic and potential energy factors (this is called enthalpy, and indicated by ΔH). From that we subtract the temperature-dependent changes (T) of the random distribution of the molecules involved (the so-called entropy [ΔS]). This can be represented by the following formula:

(1) $\Delta G = \Delta H - T\Delta S$

The kinetic energy that appears is mainly translational (E_t) and rotational (E_r). Translational energy refers to the collisions between macromolecules, which are visible as Brownian movement. For two molecules to react with each other, they must get into a specific conformation by means of rotational energy.

If they are in the correct relationship to each other, molecules can bond in two ways without reducing their potential energy: *a. Covalent bonds* These decrease the potential energy by about 100 Kcal/mol. *b. Secondary bonds* (van der Waals forces) with a decrease of only about 5 Kcal/mol.

Secondary bonds are more important in the extracellular matrix. They include:

a. Dispersive bonds. These appear with the attraction of two non-polar atoms. The reduction of potential energy is 0.05 to 0.2 Kcal/mol. Hydrophobic bonds (Coulomb forces) also belong in this category; these are very weak dispersive bonds which are generated by water. Gosline's reference (1976) gives an example of this. Apparently while elastic fibers are stretching, the reaction of water molecules with the hydrophobic groups on the elastic fibers can store elastic energy, and this energy is again given off when the fibers return to their original state. *b. Electrostatic bonds.* They can attract and repel, and thus reduce or increase the potential energy. Hydrogen bonds which mediate between two atoms as electrostatic bonds are important for the energetic functioning of the extracellular matrix.

Thus potential energy in the extracellular matrix functions either covalently (P_C), electrostatically (P_E) or dispersively (P_D). In contrast, kinetic energy primarily appears as translational (E_t) and rotational (E_r). The sum of the energy factors or enthalpy (H), of the extracellular matrix can be summarized as follows:

(2) $\quad H = P_C + P_E + P_D + E_t + E_r$
(3) $\quad \Delta H = \Delta G + T\Delta S$
$\quad\quad S = k \ln (N)$
(4) $\quad k = $ Boltzmann Constant

According to (1) and (3), the enthalpy (ΔH) of the extracellular matrix corresponds to the free energy exchange of the system (ΔG) and its amount of entropy (ΔS). The entropy is proportional to the conformation of a macromolecule and the number of spatial positions (N) which can be occupied by a small molecule such as water. Thus entropy accompanies breakdown of structures, and enthalpy accompanies the new formation of structures.

When molecules have more possible spatial arrangements or, if the volume occupied by a molecule increases (e.g., through warming), then according to (3), there will be a corresponding increase in entropy, causing the net quantity of free energy to decrease, according to (1). Reactions in the extracellular matrix can thus predominantly decrease either entropy or enthalpy. Both energetic qualities must keep nearly in balance in the organization of the extracellular matrix, which is the molecular sieve between capillaries and cell. If there is too much enthalpy, there will be too great an increase in entropy, so that the macromolecular principle of order is lost (acute illnesses, infections, allergies, rheumatic illness, neoplasms). If the energy exchange is too low, entropy is destroyed, resulting in supra-molecular states of order, which are just as incompatible with normal functioning of the extracellular matrix (sclerosis, gelosis, sarcoma).

The balance between enthalpy and entropy is extraordinarily liable to change and thus is very easily influenced on an energetic level. It is, however, the prerequisite for maintaining of any kind of molecular order in a biological flow equilibrium. In contrast to closed energy systems (Newtonian systems), the labile equilibrium of molecular reactions in an open energy system does not lead to random molecular collision, but to spontaneous, always new, interdependent states of order with self-catalytic capabilities.

Spontaneity should never be confused with chance. Spontaneity can only be present in states of order that are easily altered. On the other hand, the random molecular collisions in closed systems eventually lead to thermodynamic equilibrium (e.g., after a red-hot piece of iron has been put into a bucket of water, the temperature between the two evens out). For open energy systems, such a stable equilibrium would be the same as destruction. In other words, in living systems, "chance" does not exist. The prerequisites and maintenance of all life are bound to the spontaneity of molecular reactions in an open energy system. This is also the basis of all biorhythms. Only this spontaneity can prevent a "fatal" thermodynamic equilibrium.

Basically, all culture involves influencing the spontaneity of biomolecular interactions in the extracellular matrix through lifestyle. From the molecular point of view, the task of medicine is maintaining the individual spontaneity of molecular interactions within the framework of homeostasis.

Extremely low energetic strengths that are enough to set off spontaneous reactions in biological systems reflect a principle based on experience that has its roots in ancient medicine; that is, if highly diluted, a poison can become a healing substance. This means that if a substance can destroy the spontaneity of molecular biological reactions, it can also support this spontaneity—if its ability to react is decreased down to the level of the reactions between water and sugar biopolymers.

This principle was revitalized by Paracelsus for academic medicine. Eventually Hahnemann carried it into the 19th century in the science of homeopathy along with the motto, "similia similibus currentur" (like cures like). The "simile principle," according to which certain symptoms can be cured through a substance that can cause similar symptoms, is a treatment directive for the molecular biological principle of spontaneous reactions in the energetically open system of the extracellular matrix. Of course, cells are also an energetically open system, but they are always downstream of the extracellular matrix.

Homeopathy, unlike any other discipline of experiential healing, is oriented to the molecular interactions and transformations between sugar biopolymers and fluid-crystalline water. Since the spontaneity of these interactions is already the expression of individuality, it is easy to understand the importance of taking a detailed case history that includes biographical and social behaviors. Such detail is necessary to effectively employ the simile principle. The remedies potentized according to this principle are specific for setting off the spontaneous autocatalytic reactivity of the matrix regulation, and when applied correctly can again lead to the individual equilibrium. It is especially interesting to note that in the highest homeopathic potencies, the solute has gone through a potentizing process resulting in a medication in which the solvent (water, alcohol) contains no molecule of the solute; and this potentizing process creates a (*strukturbedingten*) structure-conditional energy content which is sufficient to ameliorate complaints and also to bring healing. From the viewpoint of molecular biology, this appropriately stimulates the spontaneity of the reactivity of the extracellular matrix. Homeopathy is thus a medical discipline geared to the stimulation and recovery, according to the simile principle, of the individual biodynamics and bioenergetics of the matrix regulation.

11. Immunological Bystander Reaction of Substantive Homeopathy

Well-established insights into the mechanism of action of low and middle potencies (D1–D14) of classical homeopathy experimentally show the so-called immunological bystander reaction (Oral Tolerance, Bystander Reaction) Heine 1987, Summary by Heine 2004, Weiner and Mayer 1996).

The immunological bystander reaction is based on low dose antigen reactions using special low and middle potency homeopathic dilutions (D1–D14) (see Fig. 14). A similar chain of events occurs whether these non-toxic preparations of animal or plant substances are administered orally, as an aerosol, subcutaneously, intravenously, or intramuscularly. They are directly phagocytized by macrophages/monocytes. Or, when mediated through mucous membrane-associated lymphocytes, these preparations are taken up by the gut mucosal M-cells, which are similar to macrophages. The dilutions are then processed by special digestive organelles (proteosomes) in the macrophage. Short sequences (5–19 amino acids) are then returned to the surface of the macrophages. These sequences are now a recognition motif for macrophages. On the surface, the motifs become bound to major histocompatability complexes, MCH (Fig. 14). The as yet non-antigen specific lymphocytes (pro-lymphocytes, Th0 cells) now recognize the motifs, and take them over by binding them to their receptors. These now "motivated" lymphocytes wander through the lymph vessels and find the nearest regional lymph nodes. There they multiply into cell clones of regulatory lymphocytes (Th3) in large enough numbers so that they can migrate to the entire organism through the bloodstream. Drawn chemotactically (e.g., through complement factors and cytokines), the Th3 cells home in on every pathological tissue region and find the lymphocytes that underlie and support local inflammatory events in those regions (pro-inflammatory lymphocytes: T4 cells and their subgroups, the T helper 1 and T helper 2 cells). The cell membranes of these pro-inflammatory cells carry not only antigenic components of toxic substances, but also also antigenic components of the pathological organ or tissue itself.

Because of the many components of the homeopathic pharmacopia (D1–D14), there will always be one or more Th3 clones with a motif similar to the persistent antigens on the cell membrane of the pro-inflammatory lymphocytes (Fig. 14). This similarity—here again the immunologic-molecular plane of the simile principle of classical homeopathy surfaces—is close enough that "motivated" Th3 cells meeting with pro-inflammatory lymphocytes immediately release the Transforming Growth Factor-beta (TGF-β). The cytokine TGF-β is quite capable of inactivating these pro-inflammatory lymphocytes (Fig. 14). It is worth noting that during the reaction, interleukin 4 and interleukin 10 are released from the Th2 cells and strengthen the efficacy of TGF-β (Chen et al., 1996, Weiner and Mayer 1996).

Autoimmune diseases are treated according to this immunological bystander principle. A low dose (1 μm/kg of body weight) of Type II chicken or porcine collagen can be given in rheumatoid arthritis to prevent joint cartilage pathology, or at least to slow it down. In multiple sclerosis, bovine myelin protein is administered, in myasthenia gravis acetylcholine receptor protein is used, etc. (summary and detailed literature by Weiner and Mayer 1996). The immunological bystander reaction can additionally be strengthened through treatment with a homeopathic preparation of the patient's own blood. The bystander reaction presupposes the existence of a ground substance that is still able to be regulated.

Immunological Bystander Reaction

(the D refers to different potentizations of substances: D4-D8 is a selection from a range of D1–D14.)

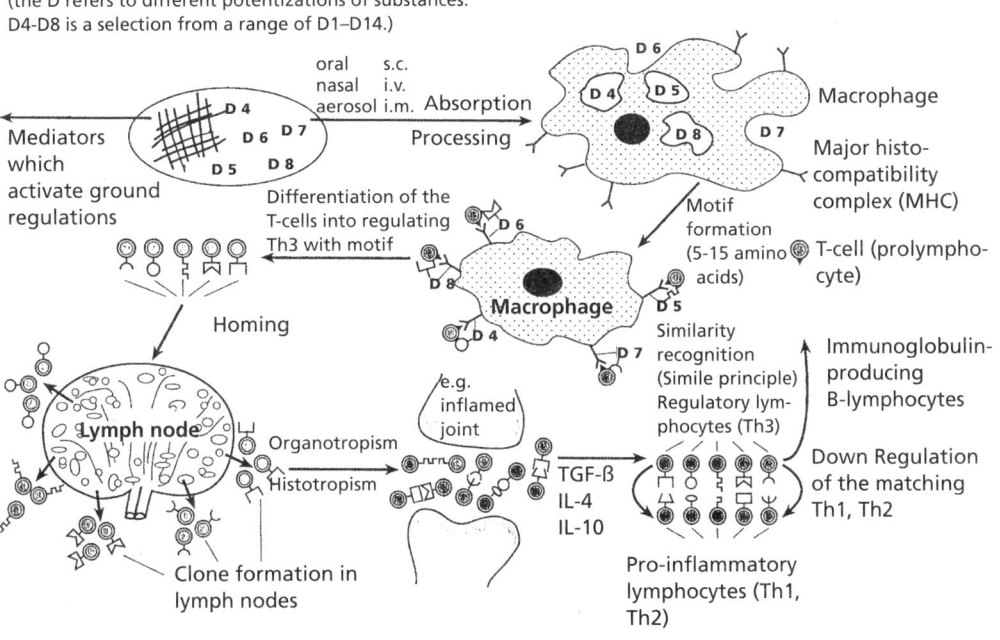

Figure 14. Process of the immunological bystander reaction. APCs = antigen presenting cells.

12. Sugar: Evidence of Pre-cellular Evolution?

Substantial traces of precellular evolution can be seen in the sugar biopolymers. Only simple oxydative steps are needed to form not only carbohydrates, but also aldehydes and carboxylic acids. The latter are changed into polysaccharides, proteins, and fats through simple chemical reactions. According to pertinent experiments in quantum chemistry done by Goldanski (1986) as well as Hoyle and Wickramasinghe (1984), it is possible for polysaccharides to develop under the extreme conditions of interstellar (cosmic) space. That is to say, in order for these kinds of reactions to occur at normal temperatures, a required energy level does not need to be reached. It can occur through quantum chemistry, i.e., the energy requirement needed for the reaction can be tunneled (Fig. 15).

Consequently, tunnel processes allow chemical reactions at extraordinarily low temperatures (Goldanski 1986). As the above researchers have shown, in this manner (*auf dieser Weise*) the documented interstellar formaldehyde formed in this manner is significant because under these extreme conditions, it can connect itself into chains of formaldehyde polymers. "Whether interstellar formaldehyde molecules could actually change into stable polysaccharides such as cellulose and starch" (Goldanski 1986) is presently under discussion. There are already indications that quantum chemistry tunnel effects can arise in biopolymers, even at normal body temperatures (Goldanski 1986).

It is strangely moving that the natural philosophical ideas of Giordano Bruno, Schelling, and Goethe about the cosmic ubiquity of the life principle are supported by the natural sciences. (In discussions with leading anthroposophic physicians, a consensus was reached that the matrix regulation system is the morphological expression of the etheric body.)

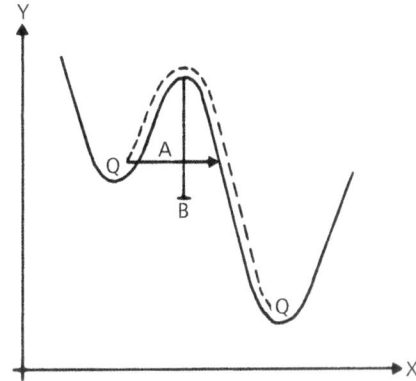

Figure 15. Quantum chemistry tunnel processes between potential troughs (Q) permit chemical reactions that could not otherwise occur. Potential energy—Y; Reaction coordinate, distance between atoms in molecule—X. The activation energy (B) is necessary to overcome the normal activation barrier (broken line). At extremely low temperatures A, the barrier can be tunneled (A) (after Goldanski 1986).

13. References

Abe, K., Saski, H., Takebayashi, K., Fuki, K., Nambu, H. *J. Interdiscpl. Cycle Res.* 9 (1978): 211. Quoted in Lemmer, 1983.

Ader, R. *Psychoneuroimmunology*. Academic Press: New York, London, 1981.

Akana, S. F., et al. "Reset of feedback in the adrenocortical system: an apparent shift in sensitivity of adrenocorticotropin to inhibition by corticosterone between morning and evening." *Endocrinology* 119 (1986): 2325–2332.

Andersson, S., et al. "Minimal surface and structure: from inorganic and metal crystals to cell membranes and biopolymers." *Chemical Reviews* 88 (1988): 221–242.

Athenstaedt, H. "Pyroelectric and piezoelectric properties of vertebrates." *Ann. N.Y. Acad. Sci.* 238 (1974): 68–110.

Baker, J. H., and B. Caterson. "The isolation and characterization of the link proteins from proteoglycan aggregates of bovine nasal cartilage." *J. Biol. Chem.* 254 (1979): 2387–2393.

Balasz, E.A., and Gibbs, P.A. "The rheological properties and biological function of hyaluronic acid," in *Chemistry and Molecular Biology of the Intercellular Matrix, Vol. 3.*, edited by E. A. Balasz. Academic Press: New York, London, 1970, 1241–1254.

Bergsmann, O., and Bergsmann, R.I. *Projektionssymptome. Reflektorishe Krankheiszeichen als Grundlage fur holistische Diagnose und Therapie*. Faculats Univesitätsverlag: Wien, 1988.

Bertalanffy, L. V. *Perspectives of General System Theory*. Braziller: New York, 1975.

Bohus, B. and S. de Wied. "Pituitary–adrenal system hormones and adaptive behavior," in *General, Comparative and Clinical Endocrinology of the Adrenal Cortex*, edited by I. C. Jonas and I.W. Henderson. Academic Press: London, 1980.

Bordeu, L. *Recherches sur le tissu muqueux ou l'organ cellulaire*. Paris, 1767.

Buddecke, E. *Grundrib der Biochemie, 2nd ed.* W. de Gruyter: Berlin, 1971.

Buttersack, R. *Latente Erkrankungen des Grundgewebes, insbesonders der serösen Häute*. Stuttgart, 1912.

Cannon, W. B. "The interrelations of emotions as suggested by recent physiological researchers." *Am. J. Physiol.* 25 (1914): 256–282.

Cannon, W. B. *The Wisdom of the Body*. W. W. Norton: New York, 1932.

Caravalho, C. R., and Vaz, N. N. "Indirect effects are independent of the way of tolerance induction." *Scand. J. Immunol.* 6 (1996): 613–618.

Cerami, A., Vlassara, H. Brownlee, M. "Glucose und Altern." *Spektrum d. Wissenschaft* (1987): 44–51.

Chiarugi, V.P., Vanucchi, S., Cella, C., Fibbi, G., Delrosso, M., Capelletti, R. "Intercellular glycosaminoglicans in normal and neoplastic tissues." *Cancer Res.* 38 (1978): 4717–4725.

Christner, J. E., M. L. Brown, and D. D. Dziewiatkowski. "Interactions of cartilage proteoglycans with hyaluronate." *J. Biol. Chem.* 254 (1979): 4624–4630.

13 ✦ References

Darwin, C. *On the origin of species by Means of Natural Selection or the Preservation of Favoured Races in the Struggle for Life.* German translation: *Die Entstehung der Arten durch natürliche Zuchtwahl.* Reclam: Stuttgart, 1963.

Demasio, A. R. *Descartes Irrtum. Fühlen, Denken und das Gehirn.* List: München, 1995.

Dillon, K. M., and M.C. Totten. "Stressful life events, personality and healthy: an inquiry into hardiness." *J. Personal. Soc. Psycol.* 37 (1979): 1–11.

Dorfman, A. "Protoglycan Biosynthesis," in *Cell Biology of the Extracellular Matrix,* 2nd ed., edited by E. D. Hay. Plenum Press: New York London, 1983, 115–138.

Dosch, J.P. *Lehrbuch der Neuraltherapie nach Huneke (Procain-Therapie), 5th ed.* Karl F. Haung Verlang: Heidelberg, 1975.

Draczynski, Th. "Die physiologische Leukozytolyse—ein Zelle–MilieuVorgang im System der Grundregulation." *Biol. Med.* 27 (1998a, b): 15–22, 67–70.

Engelberg, H. "Probable physiologic functions of heparin." *Fed. Proc.* 1 (1977): 36.

Eppinger, H. *Die Permeabilitäspathologie als die Lehre vom Krankheitsbeginn.* Springer Verlag: Wien, 1949.

Fintelmann, V. *Intuitive Medizin. Einführung in eine anthroposophisch ergänzte Medizin 2nd ed.* Hippokrates Verlag: Stuttgart, 1988.

Fischer, G. "Waschstumsdynamk und Bioenergetik." *Messen, Steuern, Regeln* (Berlin): 29 (1986): 98–100.

Franzblau, C. and B. Raris. "Elastin," in *Cell Biology of the Extracellular Matrix.* 2nd ed., edited by E. D. Hay. Plenum Press: New York, London, 1983, 65–94.

Freund, F., and Kaminer, G. *Biochemische Grundlagen der Disposition fürs Karzinom.* Springer Wien: Austria, 1925.

Fülgraff, G. "Der kontrollierte klinische Versuch—Eine kritische Würdung. *Pharmazeut. Ztg.* 130 (1985): 3309–3313.

Gautherie, M., and Ch. Gros. *Rhythmische Funktonien in biologischen Systemen.* Facultas Verlag: Wien, 1977.

Goldanski, W.I. "Quantenchemische Reaktionen bei sehr tiefen Temperaturen." *Spektrum d. Wissenschaft* 4 (1986): 62–71.

Gosline, J. M. "The physical properties of elastic tissue." *Int. Rev. Connect. Tissue Res.* 7 (1976): 211–249.

Grennough, W. T., and C. Bailey. "The anatomy of memory: convergence results across a diversity of tests." *Trends Neurol. Sci.* 11 (1988): 142–147.

Grimaud, J. A., and H. Lortat-Jacob. "Matrix receptors to cytokines: from concept to control of tissue fibrosis dynamics." *Path. Res. Pract.* 190 (1994): 883–890.

Gutmann, V., and G. Resch. "Hochpotenz und Molekularkonzept." *Therapeutikon* 4 (1988): 245–252.

Harkness, R.D. "Functional aspects of the connective tissue of skin," in *Chemistry and Molecular Biology of the Extracellular Matrix, Vol. 3.,* edited by E. A. Balasz. Academic Press: New York, London, 1970.

Hascall, V. C., and G. K. Hascall. "Proteoglycansm," in *Cell Biology of the Extracellular Matrix,* 2nd ed., edited by E. D. Hay. Plenum Press: New York, London, 1983, 39–64.

Hauss, W.H., G. Junge–Hülsing, and G. Gerlach. *Die unspezifische Mesenchymreaktion.* Thieme: Stuttgart, 1968.

Hay, E.D. "Extracellular Matrix." *J. Cell Biol.* 91 (1980): 205–223.

———. "Collagen and Embryonic Development," in *Cell Biology of the Extracellular Matrix, 2nd ed.*, edited by E. D. Hay. Plenum Press: New York, London 1991, 379–410.

Haynes, R. O. "Fibronectin and its Relation to Cellular Structure and Behavior," in *Cell Biology of the Extracellular Matrix, 2nd ed.*, edited by E. D. Hay. Plenum Press: New York, London 1991, 295–334.

Heine, H. "Vakuumprobleme bei histologischer Gefriertrocknung." *GIT Fachz. F. d. Lab. Sonderheft* Mai (1974): 531–532.

———. "Basalmembranen als Regulationssystem zwischen epithelian Zellverbänden." *Gegenbaurs Morph. Jahrb.* 132 (1986): 325–331

———. "Die Grundregulaution aus neuer Sicht. Ärztezeitschr." *f. Naturheilverf.* 28 (1987a): 909–914.

———. "Regulationsphänomene der Tumorgrundsubstanz." *Dtsch. Zschr. Onkol.* 19 (1987 b): 67–72.

———. "Anatomische Struktur der Akupunkturpunkte." *Dtchr. Zschr. Akup.* 31 (1988a): 26–30.

———. "Akupunkturtherapie—Perforationen der oberflächlichen Körperfazie durch kutane Gefäß–Nervenbündel." *Therapeutikon* 4 (1988b): 238–244.

———. "Markierung von Blutzellen mit einem Lektin–Karbohydrat–Komplex. Ertweiterte Funktions diagnostik an Blutausstrichen," quoted in *Mikrosk.–anat. Forsch.* 102 (1988c): 54–62.

———. "Grundregulation und Ganzheitsmedizin." *Natur und Ganzheithsmed.* 2 (1990a): 68–72.

———. "Grundregulation und rheumatische Formenkreis." Ärztezeitschr. f. Naturheilverf. 36 (1995): 415–426.

———. "Ganzheitsmedizin," in *Naturheeilverfahen*, edited by J. Grifka. Urban und Schwarzenberg: München 1995, 321–36.

———. *Hehrbuch der biologischen Medizin, 2nd ed.* Hippokrates Verlag: Stuttgart, 1997.

———. "Homöopathie und Immunologie. Vom Substanziellen zu Hochpotenzen." *Ärtezeitschr. f. Naturheilverf.* 45 (2004): 391–397.

——— and H. Henrich. "Reactive behavior of myocytes during long term sympathetic stimulation as compared to spontaneous hypertension." *Fol. Angiol.* 28 (1980): 22–27.

——— and G. Schaeg. "Informationssteurung in der vegetativen Peripherie." *Zschr. Hautkr.* 54 (1979): 590–597.

——— and M. Domann. "Fibronectin—plasmin–sensitive glycoprotein of the transit zone. Protection by aprotinin." *Arzeinm.–Forsch./Drug Res.* 34 (1984): 696–698.

Hewitt, A. T., H. K. Kleinmann, J. P. Pennpacker, and G. R. Martin. "Identification of an adhesive factor for chrondrocytes." *Proc. Natl. Acad. Sci.* 77 (1980): 385–388.

Hildebrandt, G. "Chronobiologistiche Untersuchungen autonomer Regulationen. Die Zeitstruktur hygiogentischer Reactionen." *Therapeutikon* 1 (1987): 70–81.

Hobbs, H. "Deletion in the gene for the low-density-lipoprotein receptor in a majority of French Canadians with familial hypercholesterolemia." *New Engl. J. Med.* 317 (1987): 734–737.

Hoeve, C. A. J., and P. J. Flory. "The elastic properties of elastin." Biopolymers 13

(1980): 677–686.

Hollemann, A.F. and F. Richter. *Lehrbuch der organischen Chemie.* W. de Gruyter: Berlin 1964, 37–41.

Hoyle, F., and E. Wickramasinghe. *From Grains to Bacteria.* University College Cardiff Press: Wales, 1984.

Huether, G., et al. "Psychische Bleastung und Neuronale Plastizität," in *Ganzheitsmedizin und Psychoneuroimmunologie. Vierter Wiener Dialog,* edited by U. Kropiunigg and A. Stacher. Facultas: Wien, 1997.

Huneke, F. *Das Sekundenphänomen, 4th ed.* Karl. F Haug Verlag: Heidelberg, 1975.

Iozzo, R. V. "Biology of Disease. Proteoglycans: Structure, Function, and Role in Neoplasia." *Lab. Invest.* 53 (1985): 337–396.

Jaques, L. B. "Herapin: an old drug with a new paradigm." *Science* 206 (1979): 528–533.

Karcher, H., and K. Polthier. "Die Geometrie von Minimalflächen." *Spektrum d. Wissenschaft* 10 (1990): 96–107.

Kefalides, N. A., R. Alper, and C. C. Clark. "Histochemistry and metabolism of basement membranes." *In. Rev. Cytol.* 61 (1979): 167–213.

Kellner, G., and G. Kleine. "Richtlinien zur Synovial zytologie," quoted in *Rheumatologie* 35 (1976): 141–153.

Köng, W., A. Bohn, K. Theobald, K. D. Brehm, and J. J. Knöller. "Die Mastzelle — zentraler Effektor bei allergischen Reacktionen." *Klinikarzt* 12 (1983): 753–776.

Kropiunigg, U. "Was ist Psychoneuroimmunologie?" in *Ganzheitsmedizin und Psychoneuroimmunologie,* edited by U. Kropiunigg and A. Stacher. Vierter Wiener Dialog. Facultas: Wien, 1997.

Lazarus, R. S. *Psychological Stress and the Coping Process.* Springer: New York, 1966,

Lazarus, R. S., and S. Folkman. *Stress, Appraisal, and Coping.* Springer: New York, 1984.

Lemmer, B. *Chronopharmakologie. Tagesrhythmen und Arzeimittelwirkung.* Wissenschaftliche Verlagsgesellschaft: Stuttgart, 1983.

Lenz, W. *Medizinische Genetik, 6th ed.* Thieme Verlag: Stuttgart, 1983.

Leupold, E. *Der Zell- und Gewebssoffwechsel als innere Krankheitsbedingung.* G. Thieme Verlag: Leipzig, 1945.

———. *Die Bedeutung des Blutchemismus besonders in Beziehung zu Tumorbildung und Tumorabbau. Il. Teil.* G Thieme Verlag: Stuttgart, 1954.

Levine, S. A., et al. "Physiological and behavioral effects of infantile stimulation." *Physiol. Behav.* 2 (1967): 55–63.

Levine, S. A., and M. P. Kidd. *Antioxidant Adaptation: Its Role in Free Radical Pathology.* Biocurrent Division: San Leandro, California, 1985.

Linsenmeyer, T. F. "Collagen," in *Cell Biology of the Extracellular Matrix, 2nd ed,* edited by E. D. Hay. Plenum Press: New York, London 1991, 7–44.

Matthews, M. B. "Connective tissue: macromolecular structure and evolution." *Mol. Biol. Biochem. Biophys.* 19 (1975): 1–318.

McBride, W. H. and J. B. Bard. "Hyaluronidase — sensitive halos around adherent cells. Their role in blocking lymphocyte-mediated phagocytosis." *J. Exp. Med.* 149 (1979) 507–515.

Mohr, H. "Das Elementare in den Wissenschaften — Möglichkeiten und Grenzen des Reduktionismus," lecture at the annual meeting Das Elementare—Bestand und Wandel, Deustschen Akademie der Naturforscher, Leopoldina (April 11–14, 1987) Nova Acta Leopold. New series (forthcoming) quoted in Peil, J., "Komplementarität von Kausalität und Finalität in der Biologie—critical review of two lectures at the 1987 annual meeting in Leopoldina, *Gegenbaurs Morph. Jahrb.* 134 (1988): 105–113.

Munck, A., P. M. Guyre, N. J. Holbrook. "Physiological functions of glucocorticoids in stress and their relation to pharmacological actions." *Endocrine Rev.* 5 (1984): 25–44.

Öbrink, B. "A study of the interactions between monomeric tropcollagen and glycosaminoglycans." *Eur. J. Biochem.* 33 (1973): 387–395.

Olsen, B. R. "Collagen Biosynthesis," in *Cell Biology of the Extracellular Matrix, 2nd ed*, edited by E. D. Hay. Plenum Press: New York, London 1991, 139–178.

Perger, F. "Die therapeutischen Konsequenzen aus des Grundregulationsforschung," A Pischinger, *Das System der Grundregulation, 8th ed.*, edited by H. Heine. Karl F. Haug Verlag: Heidelberg, 1990.

Pischinger, A. "Das Schicksal der Leukozyten." *Z. mikr.–anat. Forsch.* 63 (1957): 627–629.

———. *Das System der Grundregulation. Grundlagen fur eine ganzheitsbiologische Theorie der Medizin, 4th ed.* Karl F. Haug Verlag: Heidelberg, 1983.

Popp, F. A. *Neue Horizonte in der Medizin.* Karl F. Haug Verlag: Heidelberg, 1983.

Priebe, L. "Rythmus des Lebendigen. Thermodynamik irreversibler Prozesse—dissipative Strukturen and Lebensvorgänge." *Umschau* 81 (1981): 43–48.

Psuja, P., L. Drouet, K. Zawilska. "Binding of heparin to human endothelial cell monolayer and extracellular matrix in culture." *Thromb. Res.* 47 (1987): 469–478.

Quincke, H. *Arch. Esp. Path. Pharmak.* 31 (1893): 211. Quoted in Lemmer, 1983.

Rapoport, S. M. *Medizinische Biochemie, 5th ed.* VEB Volk und Gesundheit: Berlin 1969.

Rea, W. J. *Chemical Sensitivity, Vol.1–4.* Lewis Publishers: Boca Raton, Florida, 1995.

Ricker, G. *Pathologie als Naturwissenschaft,* Springer Verlag: Berlin. 1925.

Rindfleisch, E. v. *Elemente der Pathologie.* Thieme Verlag: Leipzig, 1869.

Robert B., Robert, L. "Aquig or connective tissues," in *Frontiers of Matrix Biology,* vol 1, edited by Robert B. and Robert L. Karger: Basel, 1973, pp.1–45.

Rokitansky, C. v. *Handbuch der pathologischen Anatomie.* Maudrich: Wien, 1846.

Schedlowski, M. *Streß, Hormone und zelluläre Immunfunktionen.* Spektrum Akadem Verlag: Heidelberg, Berlin, 1994.

Schipperges, H. *Homo patiens: Zur Geschichte des kranken Menschen.* Piper: München, Zürich, 1985.

Schmitt, F. O., J. Gross, J. H. Highberger. "Tropocollagen and the properties of fibrous collagen." *Exp. Cell Res.* suppl. 3 (1955): 326–334

Schnering v., H. G. "Die Krümmung chemischer Strukturen." *Nova acta Leopoldina NF* 65 (1991): 89–103.

Schröder, H, J. *Gesetzmäßigkeiten bei der Nekrobiose und Autolyse der weißen Blutzellen und ihre biologische Bedeutung.* Second habilitation dissertation. Med.

Fak. d. Univ.: Hamburg, 1959.)

Selye, H. *Stress Without Distress.* Y.B. Lippincott Co.: Philadelphia, New York, 1974.

Selye, H. "The Evolution of the Stress Concept: Stress and Cardiovascular Disease," in *Society, Stress, and Disease, Vol. 1: The Psychosocial Environment and Psychosomatic Disease,* edited by L. Levi. Oxford University Press: London, New York, 1971.

Sengbusch, P. v. *Molekular—und Zellbiologie.* Springer Verlag: Berlin, Heidelberg, New York, 1979.

Sherr, C. J., M. B. Taubmann, and B. Goldberg. "Isolation of a disulfide–stabilized, three–chain polypeptide fragment unique to the precursor of human collagen." *J Biol. Chem.* 248 (1973): 7033–7038.

Stux, G. *Grundlagen der Akunpunkter, 2nd ed.* Springer Verlag: Berlin, Heidelberg, New York, 1988.

Takatsuki, A., and G. Tamura. "Effect of tunicamycin on synthesis of macromolecules in cultures of chick embryo fibroblasts infected with Newcastle disease virus." *J. Antibiot.* 24 (1977): 785–794.

Thomas, F. "Die Anwendung einfacher Prinzipien der Regelung komplexer Systeme auf die Humanmedizine." *DELVR– Mitt.* 84–13. Braunschweig, 1984.

Timpl, R., G. R. Martin, P. Bruckner, G. Wick, and H. Wiedemann. "Nature of the collagenous protein in a tumor basement membrane." *Eur. J. Biochem.* 84 (1978): 43–52.

Toole, B. P. "Glycosaminoglycans in Morphogenesis," in *Cell Biology of the Extracellular Matrix, 2nd ed,* edited by E. D. Hay. Plenum Press: New York, London 1983, 259–294.

Trincher, K. *Die Gesetze der biologischen Thermodynamik.* Urban u. Schwarzenberg: Wien, München, Baltimore, 1981.

Uhlenbruck, G., H. J. Beuth, K. Oette, T. Schotten, H. L. Ko, K. Roszkow, W. Roskowski, R. Lütticken, and G. Pulverer. "Lektine und die Oragnotropie der Metastasierung." *Dtsch. Med. Wschr.* 111 (1986): 991–995.

Uno, H., et al. "Hippocamal damage associated prolonged and fetal stress in primates." *J. Neurosci.* 9 (1989): 1705–17011.

Ursin, H., and M. Olff. "The Stress Response," *in Stress: From Synapse to Syndrome,* edited by S. C. Standford and P. Salmon. Academic Press: London, 1992.

Weihe, E. "Neuropeptides in primary afferent neurons," in *The Primary Afferent Neuron,* edited by W. Zenker and W. L. Neuhuber. Plenum Publishing Corporation: New York, 1990, 127–159.

Wiener, H. L., and L. F. Mayer. "Oral tolerance: Mechanisms and applications." *Ann. N. Y. Acad. Sci.* 778 (1996): 1–453.

Wendt, L. *Die Eiweißspeicher–Krankheiten. Proteothesaurismosen.* Karl F. Haug Verlag: Hedelberg, 1984.

Wendt, L. and H. Warning. "Verschlackungs–Syndrome." *Hufeland. J.* 2 (1986): 27–39.

Wiener, N. *Kybernetik—Regelung und Nachrichtenübermittlung im Lebwesen und in der Maschine.* Econ–Verlag: Düsseldorf, 1963.

Wilner, P. "The validity of animal models of depression." *Psychopharmacology* 83 (1984): 1–16.

Yamada, K. M. "Fibronectin and Other Structural Proteins," in *Cell Biology of the Extracellular Matrix, 2nd ed,* edited by E. D. Hay. Plenum Press: New York, London, 1991, 95–114.

Part Two

1. Ground System, Regulation and Regulatory Disturbances in a Rehabilitation Practice

While the clinical picture of acute illnesses exhibits massive pathomorphologic changes and/or severe symptoms, these are absent in cases of chronic illnesses and conditions (except in cases of malignant processes and genetic defects). In these chronic syndromes, what is most noticeable are the dysfunctions and disturbances to the balanced state of health; their causes are not immediately evident. Most of the usual clinical pictures and biochemical criteria offer few or no reference points for a possible treatment protocol. This almost insures a typical misuse of medicine: a diffuse, unclear condition will be classified as the closest clinical syndrome and treated accordingly. *In this way a chronic condition is changed into an acute illness.*

The results of this flawed thinking are well known and can be seen at any time in medical units for the chronically ill, or in rehabilitation centers. The customary dosages of medications used for acute medical conditions add to the regulatory burden and cannot combat the cause of the condition. The same is even truer for surgical procedures (which are often not indicated). On the other hand, the additional iatrogenic regulatory load, in other words, medication, perpetuates and aggravates the condition. Note we are *not* questioning the validity of either the potential or actual clinical medical treatment in acute, severe diseases or in medical emergencies. We are merely emphasizing that in chronic diseases the diagnostic and therapeutic methods of clinical medicine only have limited justification, and then mainly in dealing with any acute exacerbations.

In order to achieve optimal results, the diagnosis and treatment of chronic illnesses must be oriented along the lines of biocybernetics and the conditions of regulatory pathology, even when regulation and/or dysregulation are to be examined and treated with the customary clinical methods.

1.1. Physiological Regulatory Requirements

An understanding of the problem of chronic conditions requires a brief description of certain regulatory and physiological facts.

1.1.1. The Organism: A Network System

As every living organism, the human organism is an interlinked, self-regulating system. The only way to describe such a complex, interlinked system, and as well other network systems, is through a flow of information, not merely through the flow of mass or energy. Such self-regulating systems are, as a matter of principle, capable of fluctuation.

It is rather unnecessary to mention that this human system must be regarded as a subsystem in relationship to the environment and the cosmos, and that it is made up of further subsystems.

1.1.2. The Regulatory Cycle

The smallest cybernetic unit is the regulatory cycle. In clinical medicine it is impossible to observe the function of an isolated regulatory cycle, and the following is only a basic description of the regulatory mechanisms. But there are easy-to-observe subsystems and systems that react according to the same principles. It is important to bear in mind that the cyclic design is not necessary; what is important is the *cyclic function*. This working arrangement functions to maintain the homeostasis of this closed loop, and in doing so, to make sure the least amount of energy is lost while correcting deviations triggered by disturbances. The latter corresponds to a *principle of economy*, i.e., optimally fulfilling needs with the smallest possible energy loss.

1.1.2.1. The Functional Elements

The functional elements of the regulatory cycle are the following:

1. The *sensor* that helps register the actual condition of the measured loop and transform it into signals which are relayed to the regulator.
2. In the *regulator*, the actual condition is is compared with a guideline which is usually determined from outside the cycle, and discrepancies between *target level* and *actual level* are transformed into signals, which are then relayed to the adjustor. (An example of such a signal is *a set value*.)
3. The action of the *adjustor* now corrects the substrate (underlying layer) until the actual level and the ideal level agree.

It should be mentioned that there are regulators with various functions such as differential and integral functions.

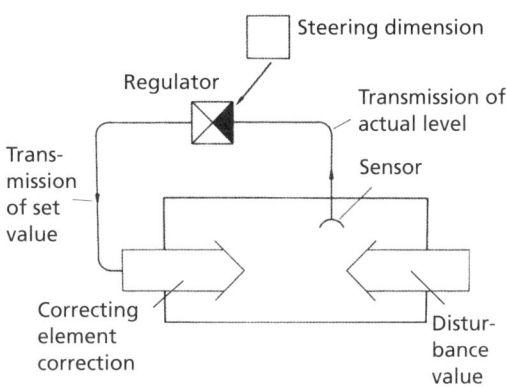

1. The regulatory cycle. In the regulator, the target level (steering dimension) is compared with the closed loop's actual level, which is picked up by the sensor. Where there are differences that exceed the tolerance limit of the system, the regulator gives the correcting element an appropriate set value (correction), and the resulting action of the correcting element brings about the correction in the substrate.

1.1.3. Type of Regulation and its Quality, and Regulatory Disturbance

The type of regulation and its quality determine the functioning of the regulatory cycle (regulatory system). Regulatory function can be observed by using one or more kinds of measurement to detect the transient response, which is defined here as the change in the system from one steady state to another steady state.

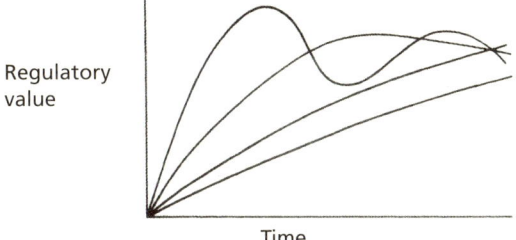

Immediate response of a system, depending on the quality of regulation.

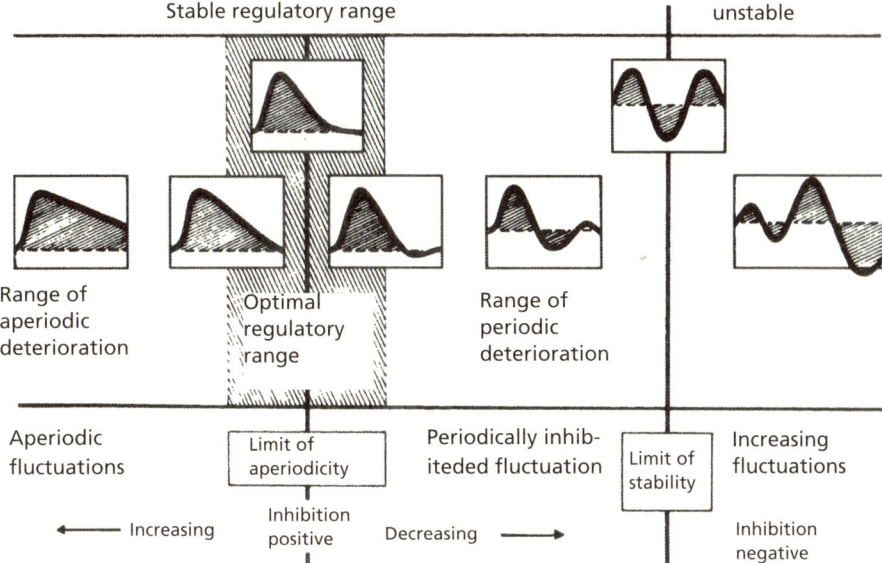

2. Fluctuating activity. Upper diagram: immediate response with adjustment of normal level. Lower diagram: leveling-out and return with short-term stimuli (acupuncture needle function). See text.

The optimal regulatory type and quality are equivalent to a moderate transient response in which the regulatory goal can be reached with the lowest energy loss in the shortest time. This corresponds to the principle of economy.

Pathological regulatory forms are the *periodic deteriorated—unstable—*transient responses, which first overshoot the limits of the regulatory goal, but eventually reach it after several after-fluctuations. The *resonance reaction* is a special type of periodic disturbance in which the after-fluctuations become ever larger, rising past the compensation level, until the entire system collapses in a *capsize reaction*. The second potential regulatory problem is the *aperiodic-inert* disturbance, in which the regulatory goal is achieved too late or not at all. Time and energy are lost through both types of disturbances. They do not correspond to a principle of economy, and since one of the primary tasks of regulation is to maintain vital functions within an optimal range, a pathological borderline area is reached.

Here it is necessary to mention a fact that has been much ignored in practice. Even at rest, an organism has no firmly set normal criteria that can be measured. Indeed, all the values swing back and forth around a middle value, so that from the regulatory point of view, there is no difference between "stable" functions and those that are recognized by their rhythmical changes.

The main causes of regulatory disturbance are defects in this system, which only refers to illnesses of the closed loop. Included here are deficits of mediators, hormones, enzymes, etc., as well as the substances that are involved in forming them. As Perger described, the blocking electrolytes that occurs through the pathological stress of a heavy metals also belong to these system defects. These deficits result mainly in aperiodic deterioration.

On the other hand, prior stress to the regulatory system from "foreign" energy leads to *overcompensation*, which in turn leads to instability of the transient response. Here the strength of the foreign energy is not as important—especially in the case of chronic diseases—as how long the effect of this energy lasts. The typical example of this is the focal process and the regulatory disturbance it creates, which correspond to the exhaustion phase in reference to *adaptation syndromes*.

In this connection, the role of the interlinking of the regulatory systems in response to stimuli must be mentioned. This will be important, particularly in evaluating regulatory tests. A disturbance variable (pathogenic in nature, or a test stimulus) that is introduced into the system can, depending on the actual functioning at various outlets (measured substrates), set off processes which have patterns of fluctuation that are completely different. Thus if there is an existing dysregulation, a test for electrolytes can show an aperiodic deterioration, while at the same time, the fluctuation patterns of the vessel walls or of leukocyte reactions can show periodically-disturbed, unstable behavior.

In each case, the deterioration of the fluctuating processes disturbs the two basic goal-oriented principles of biocybernetics—*homeostasis and economy*. The disturbed organism then functions in a completely or partially uneconomical way, so it arrives at the end of its capacity sooner. Thus the simultaneous metabolic disturbances help create degenerative changes in the organs which not only are under regulatory stress but also must perform more work.

From both the clinical and pathological points of view, it is necessary to distinguish between two different models of regulatory principles: the *stimulus reaction model* if the stimuli are only short-term (acupuncture needle function), and the *adaptation model* if the stimuli are long-term. In an initial phase of the adaptation model, the target level shifts, so the observed measured levels are higher than normal. In the later stages, this results in exhaustion of the regulatory output. In addition, the levels decrease, and the reactivity lessens, in the sense of Selye's adaptation syndrome. Many of the conditions labeled as "focal disease" can be put into this category in terms of regulatory pathology.

The purpose of this instant guide to biocybernetic facts and principles is only to show that clinical medicine's thinking in terms of syndromes, which is completely justified in acute medicine, *leads us astray in the case of chronic illness and pathological conditions.* In this case, positive diagnostic and therapeutic results can only be brought about by *thinking in terms of interrelated relationships,* and this can only be achieved through knowledge of the *regulatory and control systems* as well as their hierarchical relationships. (Cybernetic literature by Keidel 1970, Drischel 1973, Zwiener 1976, Hildebrandt 1985)

2. Chronicity as Biocybernetic Problem

As the following experimental findings and observations drawn from rehabilitation medicine demonstrate, chronic states are clinically dependent on regulatory processes and/or their dysfunctions. In exploring these, we will answer the question regarding reproducibility in this way: Concerning regulation and dysregulation, or rather eliminating dysregulation, the question still remains, since it is obvious that regulatory therapeutic intervention must be tailored to each individual because of the variety of factors that can trigger the regulatory load of each patient.

2.1. Observations in Chronic Pulmonary Tuberculosis

Although tuberculosis (TB) no longer plays as important a role as it did 20–30 years ago when these experiments were carried out, it can still be used to demonstrate typical pathogenetic and therapeutic patterns in a framework of chronicity. While considering this, it should not be forgotten for an instant that the etiology of TB infections is obvious and cannot be questioned. The problem at hand is that of *a deterioration into a chronic condition.* It must be noted that despite the intermittent epidemic character of TB, *Mycobacterium tuberculosis* is a microorganism with low human pathogenicity.

2.1.1. Pathogenic Investigation

To answer the question of how much the clinical manifestation of TB depends on secondary factors, case histories from 1542 archived medical records were evaluated according to the following factors:

1. Did a unilateral stress (trauma, surgery, organic illness, etc.) occur before the first manifestation of the tubercular lung process?
2. Was the onset of TB predominantly (ignoring minimal contralateral foci) on one side of the body?
3. To what percent were the previous stress and the onset of TB on the same side?

Results: Out of the 1542 patients, 503 met criteria 1 and 2, and of those, 75.5% developed TB on the same side as the previous stress. In 145 patients, the patients' recollections in their case histories were incontrovertibly in accord with objective, previously discovered unilateral lesions. In this group the correspondence with the previously stressed side and the side on which the TB first developed was 93.8% (Bergsmann 1963).

This investigation clearly shows that, besides the biochemical cascade whose reality cannot be doubted, there are distinct control processes that play a role in the pathogenesis of chronic disease. This discussion will be continued.

2.1.2. Therapeutic Results

With the above-described investigations as a starting point, we attempted to influence the process with therapeutic procedures that affect regulation. Neural therapy and acupuncture seemed appropriate. First, it must be established that in these experiments patients were only accepted if their sputum was permanently converted by months of optimal antibiotic therapy. They all had previously had TB residue cavities (lucency) before treatment, and after antibiotic treatment, the cavities were still there. All the patients were treated with the three-pronged tuberculostatic approach.

In each case the cavities disappeared after 3–5 months of regular regulatory therapy. One enormous cavity, documented as resistant to treatment for six years, took six months to disappear.

In patients who had urticaria in regions of the skin which were thoracic projection zones of the site of the pulmonary TB lesion, the area around the confirmed ordinary, extrapulmonary foci were flooded with anesthesia as in Störfeldtherapy (disturbance field therapy), using neural therapy. In various programs to treat pulmonary cases, acupuncture therapy, which was precisely determined by the patient's condition, showed positive results.

2.1.2.1. Breakdown of Hyper-reactive Allergy (Hyperergic) Reactions

Because the breaking down of hyper-reactive allergy (hyperergic) reactions has far-reaching applications, it is being described separately, although our initial research and findings on this subject were in the field of tuberculosis (Bergsmann et al., 1968). After we had ascertained that patients with hyper-reactive forms of tuberculosis and high tuberculosis skin reactivity inevitably had secondary disturbances (foci-disturbance fields—*Herde-Störfelder*), we took one group of patients who had high tuberculin sensitivity and regularly flooded the appropriate foci with local anesthetic before antibiotic therapy started. The result was a decrease in sensitivity by a factor of about 1,000.

The fact there was a *tuberculosis allergy* did *not* change. However, the *level of the allergic reaction* of the skin decreased.

Analogous therapeutic results were also achieved in focus-caused *hyperreactivity* of the bronchial mucosa, a finding that is frequently considered to be incipient bronchial asthma.

2.1.2.2. Increase in General Performance Capacity

In the initial phase of rehabilitation after prolonged severe illnesses, a lack of vitality or drive often makes it hard for the patient to exercise as much as needed. In such cases, therapies that affect the regulatory capacity can be a great help.

After ascertaining in individual observations that acupuncture and neural therapeutic disturbance field anesthesia could have positive results, we used ergometry (measuring work or energy expended during exercise) to check the effectiveness of these methods in rehabilitation groups of 20 patients who were between the ages of 18 and 40.

The acupuncture group was treated with a program suggested by Bishko. The neural therapy group was treated individually using local anesthetic to switch off confirmed disturbance fields.

The performance capacities of the patients were tested with pulse-regulated ergometry before and 24 hours after a one-time treatment.

Method: The ergometer brake was regulated by the pulse rate of the patient. Because of labor law regulations, we set a pulse rate during physical activity of 120 per minute. Actual performance was registered continuously during ten minutes of pedaling. There was a statistically significant increase in both the peak performance and in the overall ten-minute pedaling performance during the course of both kinds of therapy. Analysis of blood gas and concurrent respirometer readings showed that after therapy, the ability to take in oxygen increased.

For this reason, both treatment methods were adopted in the therapy routine of the rehabilitation center. Later, we used the same methods to investigate and to judge the effectiveness of the other regulatory therapeutic procedures.

2.1.2.3. Treatment of Tension Syndromes and Tension Pain Syndromes with Neural Therapy

In order to test the efficacy of the disturbance field anesthesia on the foci in a larger patient population, the following protocol was employed at the Gröbming rehabilitation center with the consent of the patients. At the conclusion of the initial examination and before any subsequent therapy session, the areas around the clinically confirmed foci were flooded with local anesthesia. After 24 hours the patients were questioned about the results, and the answer was recorded on a scale of +3 to -2 in the clinical records. A series of medical records were randomly selected from the archives and evaluated.

Number of patients	518 = 100%
After 24 hours complaint-free, or with minor residual complaints	113 = 22%
Significant improvement	156 = 30%
Total = positive results	269 = 52%

Even though the name itself "pain therapy" might encourage dramatic, hair-splitting arguments, the above numbers speak for the method without further comments. It needs to be emphasized that in this first treatment, *no topical* pain management technique with therapeutic local anesthetic was used, and that the therapeutic results were only achieved through neural therapeutic *breakdown* of the existing *dysregulation*.

However, the reduction in the need for analgesics, antirheumatics, and corticosteroids resulting from consistent use of focus disturbance field treatment in pain and function therapy in an inpatient setting, also brought about an economic effect, comparable to the cost effectiveness of focus rehabilitation described by Perger.

2.1.2.4. Regulatory Therapy for Respiration and Circulation

When regulatory therapy is successful, connective tissue turgor and muscle tone decrease and the respiratory movement changes. If, however, the respiratory movement is defective before treatment begins, flooding the disturbance field with local anesthesia will normalize the movement.

In *normal respiration*, the side of the thorax expands optimally as a result of the interaction of the "bucket handle motion" of the first rib (the action of the scalene muscles) with the expansion movement of the lower ribs (second phase of diaphragm action), and the "pump handle" motion of the sternum. During this, the manubrium remains at the same height in both inspiration and expiration.

In contrast to this, in *dysfunctional breathing*, the sternum lifts on inspiration. This causes the "bucket handle motion" to become more difficult or impossible, resulting in loss of part of the expansion of the diaphragm action. In addition, lifting the thorax during inspiration brings part of the insertion of the diaphragm closer to its own dome, and the first phase of the diaphragmatic respiratory action is lost. Since besides this, through the lifting of the thorax during inspiration, a part of the insertion of the diaphragm is brought toward its own dome, the first phase of the diaphragmatic respiratory action is lost (Bergsmann and Eder 1982).

When there is a stress on only one side of the organism, e.g., if there are unilateral foci or unilateral dysfunctions of the axial structure, dysfunctional breathing can appear only on the stressed side, while the unstressed side breathes normally. An *asymmetry* and *asynchrony* then appear in the respiratory movement.

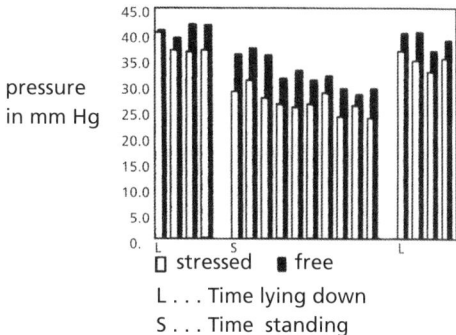

Figure 3. Schellong Test 1–Simultaneous, bilateral. During a simultaneous bilateral orthostatic test, the blood pressure amplitude on the side of the lower breath excursion, or the dysfunctional respiratory movement side, is lower than on the side with normal movement on inspiration.

During otherwise undisturbed pulmonary and thoracic function, the volume and capacity of the lungs are minimally changed or unchanged by dysfunctional movement. The resulting movement significantly impedes the work of respiration, so that more work must be expended to achieve the same inspiratory volume, which uses more oxygen. As a result, dyspnea appears at considerably lower levels of stress than if the movements were normal. At the same time, this causes the stressed patient to automatically try to do less work while breathing—he is *hypodynamic*. Respiration is not the only function actively affected by the thoracic movement during inspiration. Besides the pumping action of peripheral muscles, this movement is also the strongest centrally directed low-pressure force that affects the circulation. A disturbance of the inspiratory thoracic movement thus puts both the lungs and the circulation at a disadvantage in equal measure, and since the increase in pressure during expiration strengthens cardiac pumping, dysfunctional respiration also impedes cardiac function.

Studies analyzing circulation with Schellong orthostatic tests showed, that during a unilateral stress with corresponding unilateral respiration asymmetry on the side of the stress, the blood pressure amplitude while the patient was standing during the tests is significantly decreased.

The respiratory movement, however, does not only affect ventilation and circulation. All the internal organs are influenced by the rhythmic respiratory pressure fluctuations. On the other hand, the functional condition of the viscera, through segmental reflexes, can also affect tone and phasic action of the chest and abdominal muscles, so there is *mechano-visceral interaction*, which centers around the movement of breathing.

As our experiments (Bergsmann 1986) have shown, these relationships are significant not only in pulmonary diseases. In patients with functional disturbances caused by vertebrae other than in the thoracic spine, we ascertained a dysfunctional pattern of respiratory movement in 68.3% of the experimental and 57% of the clinically healthy subjects. Because this resulted in multiple regulatory disturbances and the loss of general performance capability, this complex can be considered to be one of the risk factors for the emergence of multiform degenerative diseases.

2.2. The Pathogenic Consequences

There is no limit to the extent to which the empirical observations and the clinical/experimental conclusions presented in the previous section can be expanded and elaborated. However, these observations, together with the principles of normal and pathological *physiology,* added to the principles of the *matrix system* by Pischinger and Heine, result in a clear picture of of the *pathogenesis* of chronic illnesses. We must, however, emphasize the difference between *etiology* (the study of origin) and *pathogenesis* (the appearance and development of an illness), since it is not the etiological factors of clinical medicine that are being questioned.

The *chronicity* factor, however, belongs to the field of pathogenesis!

2.2.1. The "Tip of the Iceberg"

In this connection, chronic diseases can be compared with an iceberg. The base—that is, the fundamental causes—is hidden at first glance. The patient chooses a doctor because of complaints which are most on his mind—mostly because of general or specific debility, general discomfort, and/or symptoms of stress or pain. These primary complaints are the visible tip of the iceberg.

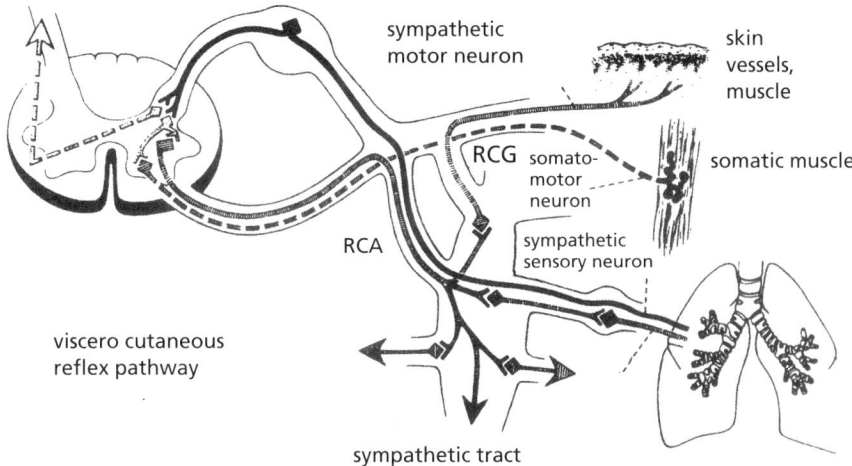

Figure 4. As an example of the viscerocutaneous reflex pathway, the functional interlinking of all substrates connected to the spinal segment is shown. The term "reflex" is misleading since the function can be altered not only by the internal organs, but also by the skin, the musculature, etc. The point is, that it functions as a network.

2.2.1.1. The Sensorimotor Control System

Ultimately, almost all symptoms related to primary complaints are sensorimotor in nature. For this reason we will briefly review, without going into every detail, the most important sensorimotor control and processing systems.

2.2.1.1.1. The Segmental Regulatory Complex

The segmental regulatory complex is the most peripheral control system. All the structures that are connected to a spinal segment are functionally related through this complex.

This interconnectedness means that when an internal organ is diseased, the signals from the antonomic pathways also relay changes in muscle tone, perspiration, etc. to the outer wall of the body. It also means that therapy to the surface of the body can change the function of the inner organs, as has been demonstrated with warm applications, poultices, and also with neural therapy and other regulatory medical treatments.

It has already been mentioned that also the autonomic nervous system is connected to the spinal segments between D2 (occasionally D1) and L2 (occasionally L3). This was, and will always be impossible to verify in purely anatomical preparations because of the intense interlinking of the peripheral ganglia and plexi. However, the segmental arrangement of the autonomic nervous system has been demonstrated in reproducible experiments.

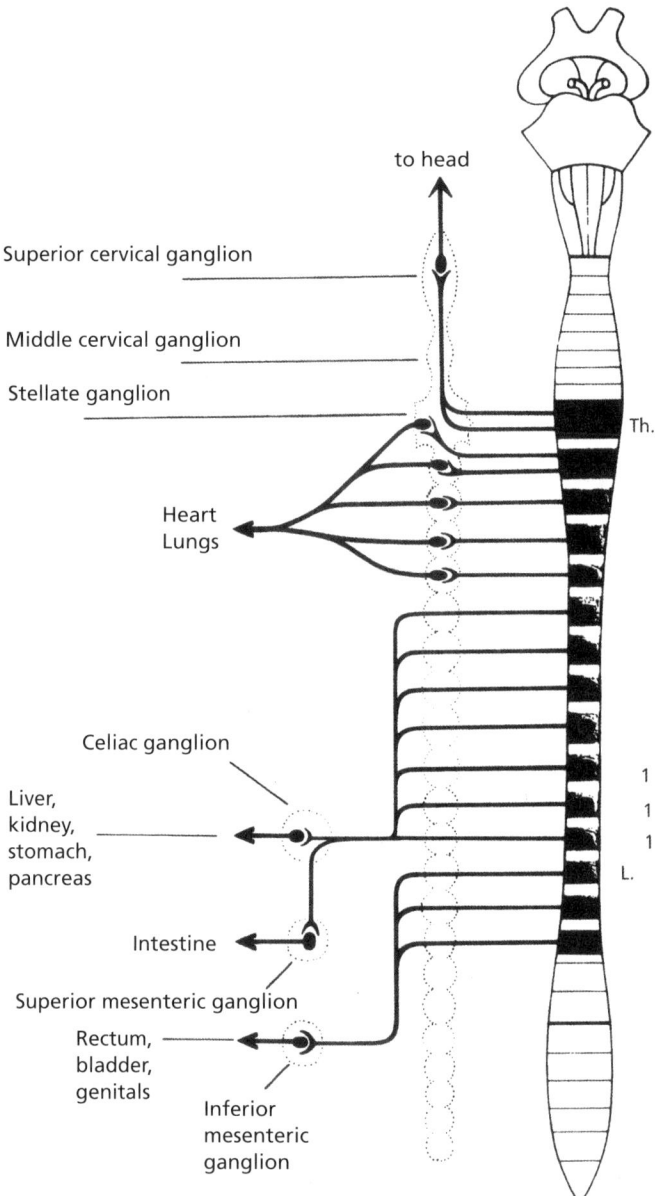

Figure 5. Segmental arrangement of the spinal cord. The segmental-regulatory complex is repeated in each spinal segment, from T2 to L2 (occasionally from T1 to L3). The sympathetic tract is also connected in segments.

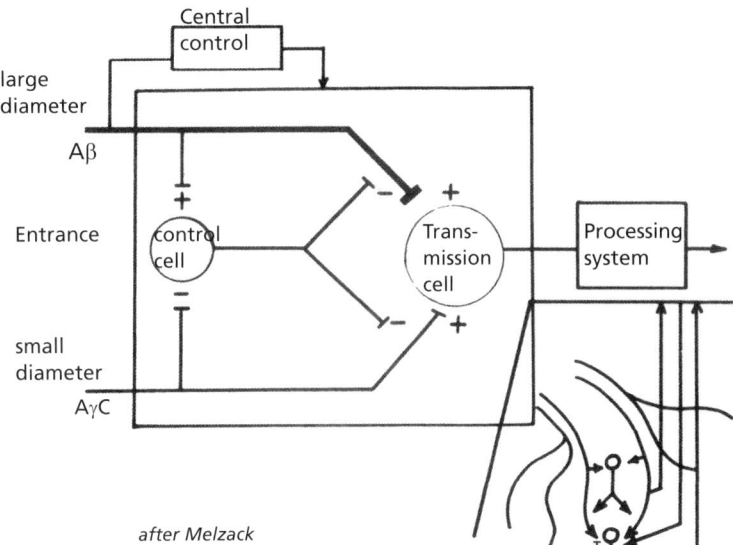

Figure 6. **The gate control system.** The posterior horn gate control itself is generally recognized, but the details are still being worked out. The system postulated by Melzack & Wall (1965) serves as a model to describe the projection processes and the mechanism of action of types of therapy that work on a regulatory level.

The entrance to the posterior horn has large- and small-diameter fibers. A control cell (SG) is connected in series with, but upstream of, the posterior horn transmitter cells; they have presynaptic feedback with both types of afferent fibers. The control cells are inhibited by afferent signals from small-diameter fibers, and stimulated by the signals from large-diameter fibers.

Results: the more information that arrives via the small-diameter fibers, the less inhibited the flow of signals is into the spinal cord. This greater flow of signals in the spinal cord can exceed the pain threshold. The greater the amount of information that comes through thick somatic fibers, the more the control cell is stimulated and the entry to the posterior horn entry is suppressed.

In the skin, each "receptive field" is innervated by dendritic branches of one sensory nerve. These dendrites overlap at the edges of the dermatomes leading to areas of heightened sensitivity, which, according to Melzack and Wall (1965) mostly correspond to acupuncture points.

The nerve endings, or other sensors, transform information from peripheral areas into signals, which are relayed to the transmission cells of the posterior horn of the spinal cord. Physiologists and clinicians have long discussed a gate control system of the posterior horn. The explanation afforded by Melzack's model is sufficient for most clinical and therapeutic problems: just before the afferent fibers synapse with a transmission cell, they receive feedback from a control cell. The control cell is inhibited by signals in small-diameter fibers (mostly autonomic) and stimulated by signals in the large-diameter fibers (mostly somatic). Stimulation means that the information input is limited and that there is an inhibition of an unlimited inflow of information.

Since the posterior horn is flooded with signals once the inhibition of input stops, the sum of the afferent signals exceeds the pain threshold. On the other hand, the quantity of information drops below the pain threshold when the posterior horn is inhibited. Acupuncture activates the somatic afferents (large-diameter fibers), and neural therapy switches off the afferent signals through the small-diameter fibers. Both can restrict the input to the posterior horn and so decrease the sensitivity to pain. The posterior horn gate control theory can also be used to interpret the effect of other forms of regulatory therapy.

Naturally, the gate control system is centrally controlled, and just as obviously, the information is transmitted and processed through the vertical hierarchy of the CNS nuclei and programs.

2.2.1.1.2. The Regulatory Control of the Musculature

Like all the other organs, the musculature is tied into the segmental-regulatory complex. Changes in tone are the regulatory response to functional or pathological information. These changes are governed by the gamma motor neuron system, which usually responds to the primary stimulus with an increase in tone.

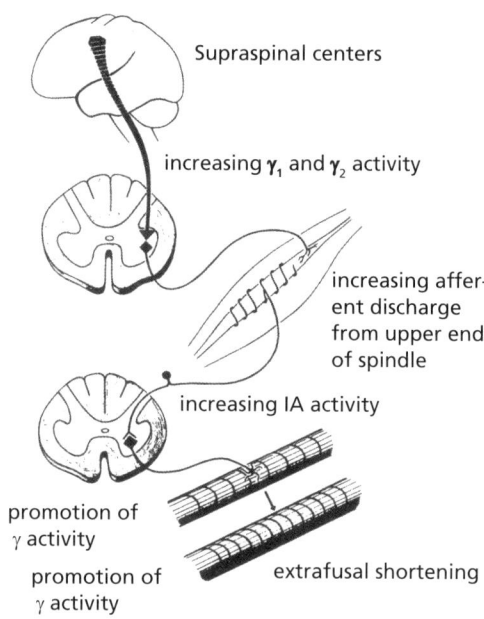

Figure 7. Gamma motor neuron. As a secondary neutral innervation for the musculature, the gamma motor neuron system is autonomous for position in space and posture (antigravity effect), and is responsible for muscle tone. Besides this, it functions as a servosystem for the alpha motor neurons. The central control is through the cerebellum (antigravitation) and through the reticular formation (interlinking and adjustment with autonomic functions). The spindle cells are stretch receptors, and as such are responsible for regulating the local tension of the muscles. The sensitivity of the spindle cells, however, is controlled by the centers.

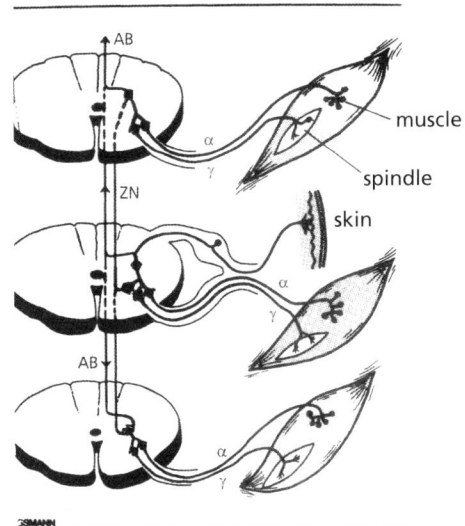

Figure 8. Trans-segmental muscle circuits. Movement complexes must be able to be pre-programmed automatically. The basis of these programs is partly from the innervation of the participating muscles from the same spinal segments, but the main part of the programs is found in the postembryonic connections in the interneuron pool of spinal cord and midbrain. The programming occurs in every motor learning process.

The following centers markedly modify muscle tone:

1. The cerebellum, where the main centers for posture and balance are situated. A second center for balance is assumed to be in the cervical spinal cord.
2. The centers of the reticular formation, which, because of their close relationship to the higher-ranking autonomic centers, balance the muscle tone with the autonomic functions that maintain life.
3. The segmental reflex complex, where the local muscle tone is adjusted to the general function of the segment.

In order to be able to carry out a *complex series of movements automatically*, programs are laid down during postembryonic development in the interneuron pool of the spinal cord and in the higher centers. This happens in all learning processes—from an upright gait, speaking, or writing, all the way to handicrafts or sports training. As a consequence of this programming, no single muscle can be activated without activating the entire kinetic chain. The differentiation of complex motion in reference to gross movement, strength, etc., occurs through spinal-segmental inhibitory mechanisms.

The system that regulates muscle tone is also part of this kinetic chain control. As a result of this, if the reflex muscle tone of a segment is stimulated, primarily the muscles innervated by that particular segmental level become hypertonic; but since they are connected to a kinetic chain, muscles in the entire chain become hypertonic. The result is a hypertonicity that spills over into neighboring segments, and muscle tension and pain in areas that correspond to the course of the kinetic chain. In clinical practice, these kinetic chain symptoms have diagnostic and therapeutic relevance in so far as:

1. The segmental-regulatory functional changes triggered through the kinetic chain symptoms can confuse the symptom picture by spilling over into neighboring segments.
2. But after tracing the symptomatology from peripheral to proximal, e.g.,

through palpation, the symptom chain leads to the trigger point.
3. That by adding peripheral therapy, a distant effect on the trigger point can be achieved.
4. These radiating hypertonic-pain symptoms are the most frequently encountered pains in everyday medical practice.

A further aspect is the parallel between these stress symptoms and the musculo-tendinous meridians of acupuncture.

2.2.1.1.3. The System of the Muscular Maximum Point

Reflex muscular hypertonicity can exceed the pain threshold to varying degrees, depending on individual differences. However, even when the tensions are minimal and nonpainful, maximal points (geloses, trigger points, etc.,) can be found in characteristic places. When gently palpating them, these points are evident as shallow depressions in the skin with a smooth surface. Using a firmer pressure on a tense muscle reveals indurations of various sizes. This triggers pain not only in the muscle, but also radiating pains in a reference zone, which lies along the course of the kinetic chain the muscle belongs to.

The significance of these maximum points for diagnosis and therapy is merely mentioned here. For more information on this system, see Brugger (1980), Travell and Simons (1983) and Bergsmann and Bergsmann (1988). According to our own observations and electromyographic experiments (Bergsmann 1994), effective acupuncture points always correspond to maximum points.

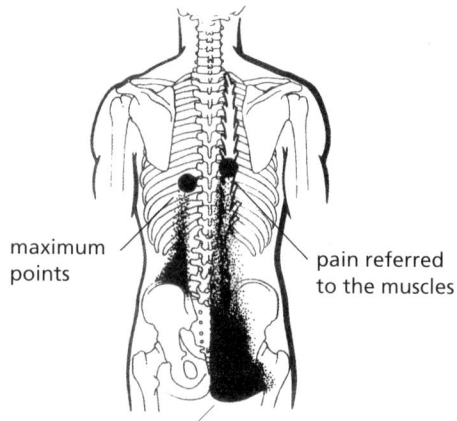

Figure 9. Example of stress symptom. The muscular nocioreceptive reflex hypertonicity follows, as the alpha motor neuron movement innervation, the course of the kinetic chain. Pain symptoms arise which spill past the boundaries of segments (pseudoradicular). In tense muscles, the maximum points are in characteristic locations.

Knowledge of this system is essential for the functional diagnosis as well as regulatory therapy of the motor system.

Palpation is an almost unrivaled diagnostic tool, even though these points can also be found by changes in skin potential, electrical resistance, and conductance, as well as changes in infrared radiation.

2.2.1.1.4. The Role of the Axial Structure in Regulatory Events

Disregarded is the fact that the spine is not only the center of postural and movement mechanics but also has a multitude of connections with the pathogenesis of regulatory disturbances and degenerative conditions, since on the one hand, every stimulus condition set off by segmental regulation leads to metabolic disturbance, and so favors degenerative changes. On the other hand, if there is simultaneous (usually unilateral) increased muscle tone, this means there is mechanical stress on the axial structure (*Achsenorgan*).[3] This leads to a lasting disturbance of posture and dynamic movement, as well as development of functional disturbances in the movement segments. Degeneration as well as functional disturbances are secondary sources of stimulation; they feed back into the segmental-regulatory complex, and so can lead to a vicious cycle and perpetuate the regulatory disturbance. Additionally, through involvement of the spinal column, a functional disturbance that is primarily unilateral can also be detected on the contralateral side, such as when additional dysfunctions arise during rotational processes in ligaments, fascia, and joints. The key point about the axial structure is that its mobile segments can be disturbed from practically everywhere [in the body], and also that they themselves can adversely effect the functions of the organism.

[3.] The axial structure is the spinal column including its joint connections, ligaments, back muscles, and the nervous system elements, together with the cranial bones

2.2.1.1.5. Spinal Afferents and the Higher Order Control System

This is not the place to repeat the complete anatomy and functional anatomy of the brainstem and the higher centers. However, we do not wish to give the impression that the segmental-regulatory complex alone is involved in pathogenesis, symptoms, and treatment. Information is relayed along the well-known afferent pathways to the higher centers, and is further processed there.

Next to the segmental-regulatory complex, the rhombencephalon-spinal control system is the most important center for developing segmental reflexes, and it is there that the different segmental regulatory functions are controlled, and the visceromotor as well as the somatomotor reflexes are channeled and controlled.

The rhombo-mesencephalon control systems contain the reticular formation, whose main task is to stabilize the inner environment through the close relationship between the autonomic centers and the gamma motor system. As a result, the reticular formation coordinates autonomic and somatic functions.

The hypothalamus, which is the control system of the diencephalon-pituitary, also helps coordinate the autonomic and somatic functions, since on one hand it is pivotal in defense against inner and outer stressors and, on the other hand, has a prime role in triggering behavioral patterns and moods.

In the limbic system (paleocortex) all the afferent signals are processed into signal complexes, and with the information from the hypothalamus, adjusted in response to the environment.

The neocortical control system processes all the external inputs and coordinates them with information from the autonomic and somatic centers. Discussion of the cortical contribution in the sense of awareness, spirit, etc., would go beyond the scope of this work.

2.2.2. The "Base of the Iceberg"

2.2.2.1. The Ground (Matrix) System

The patient's chief symptoms, which the practitioner focuses on because they approach his senses, have their basis in histochemical and biochemical functional disturbances whose point of origin is the function and dysfunction of the *matrix system*, which is the subject of this book. At this point there is no need to further discuss structure and function. To understand the central role the matrix function plays in the pathogenesis of chronic illnesses and degenerative conditions, it must be put on record that not only is it true that the matrix system is the "starter" for information to the cell, the humoral system and the nervous system, but also that the function of the matrix system itself can be changed by all functional disturbances of the tissue. This concept is most important in regards to multiple feedback mechanisms, and without it, chronicity and degeneration cannot be understood.

In this connection, the time factor plays an important role. It is impossible to explain or understand the effects of long-lasting minimal stresses (foci, disturbance fields) without considering the factor of time.

In open and functionally viable biological systems, each impulse from a short-term stimulus leads to partial depolarization of the proteoglycans, which immediately repolarizes. In contrast, minimal long-lasting stimuli resulting from localized inflammatory foci create lasting depolarization processes, which in the end must lead to structural changes in the entire matrix system. This result of this degenerative process is transformation in the direction of a gel. This means that the matrix change tends towards biological inactivity, as in the case of every other colloid that loses surface charge.

Regarding the effect of stimulation on the fields of crystalline water that lie between the molecule filaments, at present there are no experimental findings which allow us to draw a direct or analogous conclusion; however, it can be assumed that they change or completely lose their polarization, structure, and arrangement as a result of changes in electrical charges in their environment, and that this causes more of the structure of the matrix system to disappear. In this connection, the experiments of Trincher (1990) must be mentioned, that the proportion of crystalline water decreases with increased warmth, and in his opinion, life is not possible without crystalline water.

The functioning of the matrix system controls the metabolism and biological activity of the cell, according to Pischinger and Heine, and the cell generates electromagnetic information. According to the calculations of Fröhlich (1984), a field strength of 100,000 V/cm can be expected, given a membrane thickness of 0.0000006 cm and a difference in potential of 0.1 V. These high, unstable field strengths cause oscillation of the membranes and the dipoles present in them. According to Fröhlich, a resonance oscillation in the range of microwaves (about 1 TeraHz) results through the relationship of the speed of sound and membrane thickness.

More recent experiments by Popp (1984) and Klima (1981, 1987) have shown that exactly those electromagnetic oscillations that exist in coherent light, present an information system that all organisms have, and which has been scarcely noticed until now. In this information system, the switching over of oxygen from the excited single state into the molecular triplet state is recognized as a source of laser with a wavelength of 634 nm. This system appears to be significant precisely in the range of cell regeneration. According to his own laser research (Bergsmann 1994), this system is also capable of setting off resonance- and dampening-phenomena in the lower frequency range (ELF).

2.2.2.2. Information Transmission

It seems certain that these changes primarily affect the immediate environment because of the electrolability and the coherence of the matrix system. Only if the changes persist do they affect the entire organism and, in so doing, change its overall regulatory functioning. In addition, every dysfunctional regulatory process creates conditions that further deteriorate the regulatory base. It must again be mentioned that this not only affects neural regulatory processes but also local tissue and humoral regulatory systems, as well as the interaction of the various regulatory systems in terms of the network of the entire organism.

The dysfunction of the matrix system certainly does not spread instantaneously, but occurs with feedback from neural and humoral systems. In this regard one could allude to the term "compartments," currently used in clinical speech. To further support (the use of this term), serous membranes, septa, and fascia all have insulating qualities which limit the spread of information by shifting the electrical charge. But according to Nordenström (1983), the extension of biological connections via lymph vessels, arteries, and veins can once again bridge these limitations, in terms of biologically closed electric circuits (BCEC).

However, in order for the matrix system structures to be electrolabile and able to fluctuate, the matrix must also be sensitive to environmental electrostatic and electromagnetic influences, such as static fields, air electrical charges, and electromagnetic impulse fields, the so-called "spherics." This is all the more so, as Heine (1988b)demonstrated, that the matrix system is stretched from the depths of the body towards the surface in the form of cylinders which enclose the perforating nerve and vessel bundles. Since this "Heine cylinder" is ensheathed by a membrane which has little conductivity, we are dealing with an organ that not only can take in mechanical qualities and rearrange them, but is also able to fluctuate in the sense of electromagnetism, and consequently we expect that it can be identified as an organ of perception for electromagnetic and magnetic variables. This would further increase our understanding of problems relating to sensitivity to weather and sensitivity to long-term minimal electric and magnetic stress.

On the other hand, the existence of the Heine cylinders also explains how long-lasting changes to organic regulatory processes can be achieved with completely different techniques, such as massage, magnetic fields, electric fields at ELFs, puncture therapy, local anesthetic, and with laser that is introduced into the organism at acupuncture points.

Athenstaedt (1974) discovered a system of pyroelectric and piezoelectric chains that must be mentioned in this connection. According to him, the whole organism is permeated by piezoelectric dipole molecules whose polarity is arranged in similar directions. These piezoelectric chain systems consist in large part, of structural glycoproteins whose single molecules are piezoelectric dipoles In toto, however, they are capable of fluctuating due to their spiral structure. When there are dysfunctions of the proteoglycans of the matrix system, fibroblasts alter their molecular structure. This creates conditions that set more degenerative processes into motion, and it is clear that these pathological variations have fluctuation and resonance qualities that are different than those of normal glycoproteins.

On the other hand, dipole molecules which are oriented in various directions can become aligned in the same direction by relatively weak external and internal fields, in this way forming electrets, i.e., permanently polarized materials that have the same qualities as the primary piezoelectric chains. Athenstaedt further ascertained that—in dry preparations—the positive pole of the piezoelectric chains always points in the direction of growth. However, the piezoelectric function of the structural glycoproteins seems to be especially important for the basic information system of the organism: during pressure changes, on the one hand, piezoelectric qualities cause electric fields to arise, and on the other, electromagnetic fields of a specific resonance frequency are transformed into mechanical fluctuations. The latter lead to strong, periodic pressure changes in the environment that cannot continue without affecting both the metabolic current that passes through the molecular sieve of the matrix, and the actual structure of the water domains. The modulating effect of laser frequencies on the frequency of muscular activity has been proven (Bergsmann 1994).

In this intermeshed system one must always remember, given such a multitude of information systems, that there is an uninterrupted information exchange between them and multiple potentials for feedback. The author knows of no system of diagnosis that only addresses one of these systems by itself.

In practice, the evaluation of the functional state of the ground with humoral parameters is difficult. Here, the newer morphological findings of point diagnosis indicate a solution. Details are found in chapter 17, "The Point: The Window on the Matrix System."

2.2.3. Regulatory Disintegration

Regulatory disintegration as a pathocybernetic super-concept has shown itself to be particularly fruitful both in the framework of pathogenetic research as well as practical regulatory diagnosis, since it explains regulatory diagnostic phenomena, processes of the development of illnesses, and tissue degeneration.

Initially there is a local regulatory disturbance, which affects the corresponding dermatome, myotome, etc. by interaction with the segmental-regulatory connection. This disturbance also alters the vessel motion and other autonomic functions in linked quadrants via the autonomic nervous system to such an extent that the total metabolism and its regulatory processes—as well as the basic information systems—function differently compared to the undisturbed areas of the organism. The difference affects not only the level of the diagnostic measurements, but also in their temporal shifts. With increasingly strong stimuli, and as the central regulatory processes are included, this primarily local phenomenon develops into a regulatory unilateral pathology. This usually only results in a general illness at a late stage, and only under the influence of secondary and tertiary factors.

In the framework of *degenerative deterioration*, the disparity between musculotendinous tension states and disturbed vasomotility triggers degeneration of the musculoskeletal system, and even here the disturbed matrix system functions as a regulatory interface.

In one clinical work it is impossible to discuss all the physiological details in this connection. However, according to the available material, the functional disturbance of the matrix system and the resulting deterioration of all fluctuating processes, as well as the disordered interactions of the subsystems in the cross-linked network of the macro organism, comprise the *base of the iceberg.*

2.2.4. Minimal, Chronic, Persistent Stresses (Foci, Disturbance Fields)

The regulatory systems and its pathways that have been outlined here are not only stressed and disturbed by clinical disease pictures. These systems even store information from minimal chronic inflammations which can only be detected with with diligence and meticulous diagnostics because there are no or few local symptoms. In view of the limited extent and the lack of activity of these processes, most of these chronic inflammations are under threshold and do not set off any symptoms in terms of a stimulation-reaction pattern. However, since they are undetected, they release the information often over years and decades, and place all regulatory systems (cellular, tissue, humoral, and neural) under stress. The result is usually primary periodic deterioration and also instability of the transient processes, in whose wake all additional stimuli are answered with excessive reactions. In this way, trivial non-pathogenic stimuli sometimes become pathogenic.

This instability primarily affects the segmental-regulatory complex which is linked to the focus. However, since the nutrition and clearance are also changed, this results in a degenerated metabolism, that is mainly seen in the musculoskeletal system where the degenerating metabolism, together with multiple movement processes, form a *locus minoris resistentiae*. However, other organs are put at a metabolic disadvantage as well.

As a further result, the regulatory changes spread further, giving rise to phenomenon of regulatory disintegration. The regulatory pathology crosses to the other side of the body, which as previously mentioned, points to secondary involvement of the axial organs.

If this now unstable system receives a minor stimulus or gets a secondary noxious exposure, there will be an inadequate and excessive response to this "second blow," which will set off a distant disturbance whose localization is determined by the point of attack of the second blow. All further stimuli of any strength will can be considered to be secondary blows, even secondary foci. A focal illness with only a single focus is extremely rare.

However, the findings of Selye (1971) regarding the adaptation syndrome are also valid for a stress load of minimal duration. According to Selye, during continuous stress (adaptation model), exhaustion with a rigid reaction appears after a temporary tendency to overreact.

3. Diagnostic Phenomena

A system of clinical regulatory diagnosis must contain the following three criteria:

1. *Localization diagnosis* to objectify a local functional disturbance; pain must carry weight in this. However, it is much more important to use this diagnostic method to find the point of origin of a functional disturbance.
2. *Qualitative diagnosis* pertains to the disturbance of the fluctuation processes, and especially to the answer of the question, "Which fluctuation process is disturbed, and how?" This can help to determine an eventual therapeutic plan.
3. *Localization* of the *origin* of the disturbance. From the viewpoint of initiating therapy, this factor is absolutely necessary, since a causal therapy can only be initiated at the location of the "starter" of a functional disturbance.

The localization diagnosis itself, as well as the actual functional disturbance and the trigger of the disturbance serve to determine the *regulatory disintegration*. The various regulatory qualities usually differentiate between these two phenomena.

3.1. Diagnostic Criteria

Both manual as well as technical diagnostic methods follow the normal specified criteria which have already been mentioned.

3.1.1. Colloidal State

The colloidal state of the skin and subcutaneous tissue helps determine the diagnosis along with palpation and electrical measurements. Since this colloidal state depends on the perfusion of the capillary bed, it can also be correlated to thermodiagnostic findings.

The engorgement, as well as the shrinking of the tissue colloid, are regulated by autonomically controlled functional changes of arterio-venous anastamoses (AVA) and by the tone of the smooth muscles of the skin which also are autonomically controlled. The AVA are rhythmically working connections between high pressure and low pressure systems of the circulation, in which most anastamoses consist of three to four vessels. When there is a reflex stimulation of an area, it is primarily the pre and post-capillary sphincters that increase in tone, so that the capillary bed no longer has a full current going through it. This leads to hypoxia and acidification of the tissue, in which changes in the pH and fluid content of the tissue shifts the colloidal state from the "sol direction" to the "gel direction." Through this, however, the dielectric constants of the tissue are changed along with all the bioelectric functions and measurable values. Even the heat radiation coefficient and the local temperature are affected. The second, and most misunderstood effect of this process is that the blood which is shunted from the arteries to the veins is mixed. Because of this mixing, the venous blood contains increased amounts of all the blood components that are intended for the use of the tissue (mostly oxygen).

3.1.2. Projection Symptoms of Internal Organs

The emergence of the projection symptoms in coordination with the sensorimotor system has already been mentioned and can be regarded as the apparent trigger of regulatory disintegration. Head (1898) and Mackenzie (1911) began to organize findings on this "reflex sign of disease," and Hansen and Schliack (1963) summarized them. More recent research, especially in the field of the musculoskeletal system has shown, however, that the problem of organ projections into skin and subcutaneous tissue is widespread, and that by including control systems that cross segmental limits, distant radiating clinical symptoms can arise. However, these symptoms can be traced back to their origins by a knowledge of the control system. These relationships were most recently presented by Gleditsch (1983) and Bergsmann and Bergsmann (1988).

3.1.3. Acupuncture Points and Meridians

The connections between sensorimotor system, kinetic chains, and acupuncture points (Bergsmann and Meng 1982, Bergsmann and Bergsmann 1988) have been mentioned above as has the enduring analogy between musculo-tendinous trigger points and acupuncture points. The discovery of the Heine cylinders further helped to demystify the acupuncture system, and allowed the empirically and experimentally verified functional diagnostic potentials of this system to be dispassionately discussed.

In terms of functional diagnostic phenomena, the fact that the Heine cylinders are supplied with vessels and nerves is significant because only this can explain the electrical and temperature differences between the point and the surrounding skin. This information even explains the many potential influences that autonomic and somatic disturbances can have on the function of the point.

As far as the diagnosis and research of the matrix system is concerned, the acupuncture point can be seen as the "window on the matrix system" (*see* chapter 17.)

3.2. Somatotopes

The somatotopes which go past the basic projection and the acupuncture system are only mentioned here and are not discussed further in the next section. Gleditsch (1983) published detailed system of the earlier known somatopes.

3.2.1. Perfusion

The autonomic nervous system directly or indirectly controls both central perfusion variables such as amplitude, flow volumes, and speed, as well as the parameters of the capillary bed. Because of this, we may draw conclusions about the autonomic status of an experimental area by using appropriate methods (*see* the section on colloidal state, *in* 16.1.1).

A direct microscopic investigation of the capillary bed is not practical for diagnosis because of time and expense; however, ground-breaking experiments were done by Brückle (1963) on the mucous membrane of the lips. The author ascertained that, on a microscopic capillary level, in the field of regulatory stress of the capillary flow, the capillary bed changes depending on the stimuli. In doing so, the changes of the flow range from acceleration to slowing, to reversal, and finally, to the exit of serum (from the blood vessels) and formation of para-vasal "Seas." The analogy these results have to Ricker's thought system in *Relations Pathology* (1924) should not be ignored.

3.2.2. Humoral Parameters and Leukocytes

The observation of leukocytes and humoral parameters following varied standardized stimuli is probably the oldest functional-analytic diagnostic method. The changes observed in this way should be evaluated according to the above-described biocybernetic criteria. As has already been mentioned, the perfusion of large vessels is also subject to regulatory disintegration; this also includes the dysfunction of the arterio-venous anastamoses in the reflexively affected sections, so that asymmetrical humoral and cellular measurements can be observed in conditions that are primarily on one side of the body.

3.2.3. Muscle Activity

When there has been tonic prestressing of a kinetic chain, adding a work load leads to more muscle activity than giving the same work load to musculature that has not been prestressed (Bergsmann 1988). According to our findings (Bergsmann 1983), electromyographic experiments can also be used to understand short-duration rhythms under various requirements, as well as to objectively understand the problem of harmonic relationships.

4. The Point: The Window on the Matrix System

Humoral regulation diagnostic procedures are mostly lengthy and time-intensive. This makes clinical matrix system research considerably more difficult, and is certainly also one of the reasons for the hesitancy in implementing regulatory diagnosis and in abandoning the methods that were already a part of clinical diagnosis half a century ago.

Here, the discovery of the morphological structure of the acupuncture point and its relationship to the matrix system has opened possibilities, and justifies the already-available diagnostic procedures that go further than the customary clinical ones.

4.1. Morphology

A histological study by Kellner (1979) of 14,000 skin sections showed that at various points (but not all), there was more than a chance collection of sensory elements, but that there were no structural differences between these points and neutral skin.

Recently Heine (1988b) discovered that underneath the points he studied, vessel-nerve bundles penetrate the fascia and bring a cylinder made of proteoglycans of the extracellular matrix with them. This cylinder is sealed off by a thicker proteoglycan layer and is bordered at its upper (outer) side, but nerve endings emerge from this upper side. Considering the problems associated with diagnosing the matrix system, these cylinders present a relatively large and concentrated part of the matrix substance primarily in close contact with the surface. This makes technical access and interpretation of measurements possible.

4.1.1. Possible Functions of the Point Organ in Acupuncture

According to Heine's description, the point organ has many functions.

- From a mechanical point of view, it is a viscoelastic system that absorbs jolts and pressure.
- The proteoglycan network that lies firmly attached to the surface of the organ is principally able to fluctuate and therefore is capable of reacting to stimuli that have electromagnetic and magnetic properties.
- A network of electro-labile molecular filaments makes up a biophysical storage system for electrical charges, namely an accumulator. The surrounding thicker layer with its insulating properties also brings to mind a condenser.
- As a result of their electrolability, proteoglycans react to every type of stimulus with depolarization, and can transmit these as chain reactions in the matrix system over great distances. This guarantees the continuity of primary information relay from the point to distant regions of the body.

- It must be considered that it is possible for muscle tension to influence the point from the area beneath the point where the matrix system and the vessel-nerve-bundle go through the gap in the fascia, since every change in tension alters the flow dynamics and the ability of the capillary bed to react in the domain of the point.
- The vascular bundle indicates the intense relationship with temperature regulation and the vasomotor system.

This gives weight to evidence for the diagnostic possibilities regarding point measurements. It is then necessary to study the question of the functional relationships between organ and acupuncture point.

4.1.2. The Question of Functional Relationships

The pertinent investigation and substantiation must take into account both directions: from the acupuncture point to the organ and from the organ to the point.

4.1.2.1. Changes in the Organ due to Stimulation of the Point.

The first verification of efficacy of acupuncture with Western medical methods (outlined by Bergsmann 1974) focused on the point Bladder 17 (B 17), which is called the "agreement point of the diaphragm, reunion of yin and yang, and the master point of blood distribution." This point affects the function of the diaphragm.

As a prerequisite, the diaphragm needed to be in a state of functional state of tension, but with no mechanical restriction. After stimulating point B 17, which lies approximately next to the spinous process of the 7th thoracic vertebra, there was an obvious relaxation of the contour of the diaphragm and an increase in the diaphragmatic amplitude.

In the first series, the diaphragmatic amplitude of over 200 patients was thoroughly examined using serial X-rays. The results were characteristic. Even random observations came up with the similar results.

A one-time treatment of point B 17 had only a slight effect on only vital capacity and resistance in the respiratory passages. However, the expiratory reserve capacity was diminished in proportion to the relaxation of the thorax—the respiratory neutral position was shifted in the direction of expiration. Corresponding to the improved flexibility of the thorax, the inspiratory potentials of the intercostal muscles significantly increased in the electromyogram.

All the compiled results, including palpation of the muscles, indicated a relaxation of the thoracic movement complex. This decreased the work of respiration and reduced subjective dyspnea.

Starting by observing relaxation, we studied the *deqi*[3] sensation of acupuncture. This is set off through deep needling in certain points that have close relationships to muscles. Electrical or mechanical input is applied to the needle, and this results in a feeling of heaviness and

---

[3.] *Deqi* is the successful stimulation of *qi* energy at an acupuncture point, often accompanied by a particular type of sensation.

warmth in relationship with hypoesthesia which spreads in the direction of the target area.

For these studies we decided to use the Triple Warmer points 5–8. These points are on the distal forearm and have a close relationship to the finger and wrist extensors, so that stimulating them activates the extensor chain of the upper extremity.

Using a digitally supported "Biofran" amplifier (Schmidt Elektronik, Lindau), an electromyogram was taken of equal length sections of both the pars horizontalis of the trapezius muscle, which is an extensor, and of the pectoralis muscle, which can be classified as a flexor.

The criteria for comparison were as follows: the mean frequency (F), the mean amplitude (A), and the product of the frequency and amplitude, which was automatically calculated and printed. Preliminary experiments also showed that the value that yielded the most information was F x A.

The results shown in Table 1 were taken from a series of eleven analogous myographs. The experimental results show that product of F x A clearly decreases in the trapezius, that is, in the affected kinetic extensor chain, while it remains unchanged in the pectoralis muscle. The experiment also shows that the effect is achieved through (peripheral) decrease of the amplitude, which is centrally regulated, remains unchanged. In *deqi*, only an increase in frequency triggered the voluntary motion needed to make a fist.

Using Western clinical methods, this verifies that the function of internal organs and muscles can be altered by stimulating certain points, and that there is a regulatory relationship from the acupuncture point in the direction of the organ.

4.1.2.2. Change in Physical Functions of the Point in Disease of the Associated Organ

The results of physical measurements on neutral skin sites and at acupuncture points show great variation in distribution both among individuals and dependent on the situation. Because of the large number of variables which clinical analysis can seldom eliminate, experimental comparison among individuals shows little promise. However, eliminating these variables is not necessary when diagnosing individual regulatory behavior. The pertinent questions are:

- What is the initial condition of the patient and/or one of his organs?
- How does the matrix system process a test stimulus?

This can allow us to judge the functioning of the matrix system.

In order to understand the functional control [*Auslenkung*] of the matrix system through a local process, the difference in measured values must be understood to be an expression of *regulatory disintegration*, which ultimately allows us to determine the status of the matrix system and its ability to react.

In the regulatory field that is unaffected, the matrix system is open, and thus able to react and regenerate. On the other hand, the patho-information emanating from the pathological process influences the matrix system, and primes one of the regulatory compartments, separated from macroorgizational regulation, to respond in one of three ways—in an excessively labile manner, or sluggishly, or with no response at all.

In patients under general stress, it is only the response to a stimulus and the subsequent functional ability of the matrix that can furnish information about the present state of the matrix system.

This variation in response to a stimulus ultimately depends on loading the structures of the extracellular matrix with electric charges. However, for verification of this we must await the results of future histochemical and histophysical experiments.

4.1.3. Phenomena of the Activated Point That Can Be Palpated

Palpation is a rapid way to perceive changes in the consistency and elasticity of the tissue, which are all physical qualities. In addition, since it is the most frequently used method in clinical practice, its results can serve as a starting point.

In an "active" point, palpable features include:

- A characteristic location in relation to muscles, tendons, periosteum, and fascia.
- In surface palpation, the point presents as a flat depression, whereas the texture of the skin is smoother.
- Skin over the point is less mobile, e.g., "the finger stays put."
  With greater pressure the muscle under the point feels noticeably harder.
- Deep in the center of the point, there is a palpable hard cord with a small nodule.
- There is a heightened painful sensitivity to pressure.
- Mechanical stimulation of the point sets off a referred pain in an associated distant reference zone, which, however, is always in the kinetic chain of the point.

These symptoms are *not* present over normal, *non*-affected muscles.

TABLE 1

	Pectoralis muscle			Trapezius muscle		
	F	A	F x A	F	A	F x A
Pretreatment values						
Hand relaxed	4.6	6.4	29.4	6.2	4.1	25.4
Fist	5.2	6.3	32.7	5.4	6.1	33.0
Deqi						
Hand relaxed	7.6	3.9	29.6	6.3	1.3	8.2
Fist	11.3	3.3	37.3	23.8	1.6	38.1

F = frequency
A = amplitude

It is important to note that regulatory processes at the point can be diagnosed just by assessing its qualities with simple palpation, since the non-affected—inactive—point is not palpable, while the activated point—regulatorily changed—can be discerned by palpation.

Since the qualities of the point qualities can change in seconds, the results of therapy can also be assessed by palpation.

4.1.4. Thermal Phenomena

We used a Medical IR (infrared) Thermometer from the Barnes Company (Conn., USA) for our thermal tests. Analogous results can also be obtained with a thermometer that is in contact with the body. We prefer the IR method because it works without physical contact, and thus without additional physical stimulation.

Steady-state experiments to compare "free" and "stressed" points. Acupuncture point Lung 8 (Lu8) was used, which is at the level of the styloid process of the radius, over the radial pulse. The subjects were patients who had severe unilateral pulmonary tuberculosis.

TABLE 2: Comparison of the infrared radiation of the point. Lu 8 (in degrees C) n = 20

	Free side	Stressed side	Difference
Inactive processes	32.45	32.0	-0.45
Active processes	31.5	31.95	+0.45
Difference	0.95 (t)	0.05 ns	0.9 s

4.1.4.1. Temperature Regulation Tests

In order to gain insight into temperature-regulating ability during various disease stages, the two hands were cooled and the reaction of the side-to-side differences in temperature—i.e., temperature-regulated disintegration—was investigated. The difference on the stressed side between the point and neutral skin, and the difference between stressed point on one hand and the free point on the other were recorded in graph form. The resulting curves indicate that the active process exhibits the maximum labile response, and the healthy process shows a minimal labile response and an inhibition of the regulatory process.

4.1.5 Electrophysiological Phenomena

4.1.5.1 Conductance investigations

After preliminary investigations with various measurement systems, which established that in chronically ill patients, there is a positive potential and increased conductance on the side of the pathological process, systematic investigations were carried out at various points with an instrument designed by Woolley-Hart (Bergsmann and Woolley-Hart 1973).

Table 2 clearly shows that the two groups of patients have the same average measurement values but are distinguished by the scattering around those mean values. However, a difference can be distinguished between free point and stressed point for the two groups with a 5% probability of error. In the active processes, the stressed point is 0.45°C warmer, and in the inactive processes 0.45°C cooler than its non-stressed counterpart.

This instrument permitted the voltage to be increased in increments of 0.5 V and made it possible to measure the resulting current in µa (microamps). There was an almost linear increase in the lower voltage stages, corresponding to Ohm's Law, up to a high voltage which varied according to the individual, after which the current began to rise at a steeper gradient or curve. This voltage was termed the breaking point, and turned out to be the most useful comparison criterion for our investigations.

In connection with this investigation, Maresch constructed a voltage ramp that rose from 0 to 12 V in 50s, and then regressed. The resulting current was graphed as a hysteresis curve. In this type of test, it was not possible to define the breaking point clearly; it could only be estimated. The best operating point was the current at 12 volts (or alternatively, also at 9 volts) and the maximal hysteresis (Bergsmann 1974). In all investigations, measurements at the activated point were compared with both the free point and neutral skin sites.

The comparative values shown in Table 3 relate to point Lung 8, which has a close relationship to the thorax, and which is found in the area of the styloid process of the radius, directly over the radial artery.

TABLE 3: I/V (CURRENT/VOLTAGE) MEASUREMENT AT LUNG 8— MEASURED IN µA

	FREE SIDE		STRESSED SIDE	
	Lu 8	Neutral skin	Lu 8	Neutral skin
Unilateral minor lung TB n=8	2.3	1.9	1.8 ←t→	1.1
	↑ s ↓	↑ s ↓		
Unilateral major lung TB n=10	0.8	0.7	1.4 ←s→	0.7
	↑——— s ———↑			

The comparisons shown in Table 4 refer to the point, Liver 9, which has a close relationship with the liver and is found at the level of the medial edge of the crease of the knee.

TABLE 4: I/V (CURRENT/VOLTAGE) MEASUREMENT AT LIVER 9, MEASURED IN µA

	LEFT SIDE		RIGHT SIDE	
	Liv. 9	Neutral skin	Liv. 9	Neutral skin
Patients with a healthy liver n=12	2.2	1.6	2.1	1.5
Patients with liver disease n=15	6.7	3.8	13.3	6.2

For those with liver disease, it is not necessary to calculate the comparison between each pair of values, since the highest value on the left is always less than the lowest value on the right.

Both investigations clearly show that the homeolateral processes measurably alter the function of the matrix at the point. The function at the point allows us to draw conclusions about the state of the body region, which is related to the point, according the principles of acupuncture.

The I/V (current/voltage) curves are well-suited to investigations on the trunk. Here too, there was a clear difference between the point and neutral skin on one side, and the point over the process and the contralateral point. At the same time we found a significant dorsal-ventral difference, corresponding to the site of a lung process which was verified by tomography.

Further studies showed that effective acupuncture is well-suited for balancing and at the same time for normalizing the measured values between the points. However, in our investigations we had to take into account that during the longer-lasting experiments, subjects' prolonged standing or sitting unavoidably created orthostatic stresses which altered the bioelectric properties of the points. We pursued this phenomenon further in a personal experimental program, and ascertained that it occurred at all points, but was particularly strong at points on the lower extremities. These events were especially obvious in legs with varicose veins. On the other hand, we verified that objective electrical measurements of a reaction to a stimulus of minimal intensity are only possible while sitting, that is, with reduced orthostasis.

Recent investigations carried out with a modern, completely electronic measuring instrument, the Impulse-Dermo-Test, confirmed our previous findings.

4.1.5.2. Experiments of Potential Differences

We consider the primary bioelectric variables to be the local potentials and the resulting differences in potential. These variables are also to be regarded as a connection between biophysics and biochemistry, in so far as every chemical process also changes the relationships of the local bioelectrical charges. Variations or perfusion will alter the current potentials. Physico-chemical influences vary membrane potentials, and every stimulus is accompanied by metabolic changes that lead to local shifts in potential. External electrical and magnetic fields align dipolar molecules, whose polarization also influences the local potentials. All of these charges plus a few more, considered in total, are part of the measurements of the local potential, and as a result, determine the differences in potential. The magnitude of the fields resulting from these potentials leads to charge displacements and ion migration in the matrix system, and determines its functional state. Seen in this way, differences in potential are the pivotal points of the physico-chemical events.

We always found positive potentials compared with the contralateral side, both in the palms of the hands as well as at point Lung 8 on the same side that had the more severe lung processes. In comparison, during indurative processes, the side the process was on had a negative charge in comparison to the contralateral side.

We recently investigated this phenomenon with the previously mentioned Impulse Dermo-Test instrument. Using this method, the potential differences between the electrodes are recorded 4 times, after which the conductance value is recorded between every two potential measurements. In this way, a raw potential and its alterations are recorded through the charge flow and the conductance (I/V) measurement. Since every current introduced from outside the body alters the bioelectrical status of the measurement site, this method can be considered a standard test of bioelectrical regulation.

The following results were obtained from 23 subjects with subclinical, but palpable tension in the upper and lower arm. The measurement was taken at the trigger point of the extensor carpi radialis muscle, which corresponds to acupuncture point Large Intestine 10. Table 5 shows that the trigger point of a subclinically tensed muscle has a lower potential than the its relaxed counterpart.

TABLE 5: EXTENSOR CARPI RADIALIS MUSCLE

DIFFERENCE IN POTENTIAL	STRESSED ARM	FREE ARM	DIFFERENCE
Mean (mV) in mV	104 mV	119.34 mV	14.78
st. dev	79.00	64.12	8.7

We must emphasize here that the only subjects we examined were those who had subclinical tension but felt no tension or pain.

In the following, rehabilitation patients with one-sided tension pain were studied, and the potentials at the painful acupuncture points in the tension area were compared with those at the symmetrical but inactive point on the other side. Of course, the position of the counterelectrode was not changed. Table 6 shows the results of 19 double measurements.

TABLE 6: POINT MEASUREMENT— POTENTIALS

ACTIVE POINTS	N = 19	INACTIVE POINTS	
P1–95 +,	–67 mV	– 58 +,	–48 mV
P2–134 +,	–48 mV	–107 +,	–65 mV
P3–130 +,	–53 mV	–129 +,	–63 mV
P4–134 +,	–62 mV	–152 +,	–60 mV

Paired comparison for P1
D (average value) = 37
sD (standard deviation) = 17.64
$t = 2.1206$, $p > 0.05$

Besides having significantly lower primary potential, it is obvious that the pain points have potentials that change less after the electrical stimulus conductance measurement than do those at control points.

Comparative electromyography studies using standardized isometric stress have shown that the product of the frequency times the amplitude is significantly higher in tensed muscle than in muscle with normal tone. This leads to the question of whether the potentials of the points which lie directly over the muscles are altered directly by the action potentials of the musculature, or whether they are altered by tension states in the fascia and the resulting change in perfusion in the area of the point.

With these results, conclusions can be drawn regarding the local state of functioning of the matrix. This mainly indicates both depolarization processes and a reduction of the biophysical (and biochemical) capacity to react. Both of these are symptoms of regulatory disintegration.

4.1.5.3. Harmonization of the Rhythm

Recently, we found a previously unknown effect of point stimulation. A point in a muscle group that has a functional relationship with that muscle group, according to acupuncture principles, was needled and treated with a laser. As a result, a frequency spectrogram that was previously disarranged was brought into harmony.

4.1.5.4. Equipment and Methods

A digitally supported 2-channel amplifier system from Schmidt Elecktronik (Lindau) was used. The computer program processes the raw potentials of the muscles registered by surface electrodes, and gives the average frequency, amplitude, and the product of the frequency times the amplitude. Beyond that, it also furnishes a Fourier analysis of 1–33 Hz.

Using a standard isometric stress, myography was performed in area of tension on patients with a variety of tension and pain symptoms. The investigation was repeated during and after treatment with a variety of therapeutic methods.

4.1.5.5. Results

In connection with the theme of this chapter, there were interesting changes in the Fourier frequency spectrogram brought about by acupuncture and laser therapy. While before the therapy there was no apparent order in the sequence of maximal and minimal frequencies, during and after the therapy, wave-shaped sequences in 4–7 Hz increments were observed. *This corresponds to the integer harmonic relationship of the overtone series.*

This phenomenon was not only observed in treating points situated near the symptom area, but was also observed in irradiating or needling points some distance away from the symptoms (e.g., the ankle in shoulder symptoms). However, these results were not achieved by treating points that had no relationship, according to acupuncture principles, to the target area, nor by treating areas of neutral skin.

This experiment shows that stimulating a point with two different qualities brings order to the rhythmic relationships of the organism. Even if the participation of genetic rhythmic centers is beyond question, it must be assumed that, as a structure capable of oscillation, the matrix system participates in perceiving and transmitting frequencies. It is also probable that the matrix system is capable of changing mechanical stimuli into oscillations.

The appearance of an integral harmonic arranging under the influence of acupuncture or treatment with electromagnetic fields (lasers) indicates a problem that has been generally ignored in medicine up to now—the integral harmonic arranging of the body's own rhythms, which according to Hildebrandt (1985), have this integral relationship. Our investigations show that this rearranging is also observed in the combined frequencies of somatic muscle. Pathological hypertonicity seems to interrupt this arranging ability and effective therapy seems to restore it.

4.1.6. Synopsis

In the sense of an ongoing research into the clinical and practical aspects of the matrix system research, the acupuncture point can be regarded as the window to the matrix system.

A series of palpatory, temperature, and electrical measurements on a variety of points have shown that recording these measurements at the point can provide good insights into the functional state and the regulatory potency of the matrix system. In this connection, recording differences in electrical potential is significant, since conclusions can be drawn from them regarding the molecular charge load.

The relationships to the origin of rhythms need further studies.

5. Diagnostic Methods

We begin this chapter with an emphatic warning! *All* the methods described here give *reflex-regulatory* results only. They cannot show patho-morphology which may eventually manifest; they can only show *functional states*. They therefore must be regarded as screening tools or as complements to normal clinical diagnosis done with imaging methods and biochemical criteria.

5.1. Palpation Reflex Signs of Illness

This is not about the usual palpation of pathomorphology in clinical medicine, but rather about ascertaining reflex changes in the skin, subcutaneous tissue, and muscles. In trying to feel the reflex-activated areas, one must palpate for the different levels with various amounts of pressure, but always with the lightest touch possible. Knowing the dermatome innervation and the kinetic chains makes it easier to find the correlation with the specific organs. The system of reflex signs of illness of internal organs as well as that of the relationship between the organs and the musculoskeletal system is well known (Bergsmann and Bergsmann 1988), but describing them is beyond the scope of this book.

Palpation is the most practical diagnostic method used in everyday medical practice. In a comparison, the results of palpation differed less than 5% from instrument-supported findings. An experienced practitioner can also use palpation in diagnosing focal lesions, since the least amount of the continuous stress of a focus (e.g., tooth or tonsil), just like other processes, leads to palpable changes in the corresponding reflex zones.

5.1.1. Thermodiagnosis and Infrared Diagnosis

Thermodiagnosis is one of the oldest methods of medical examination. The infrared radiation diagnostic methods developed in recent decades, the development of the thermistor sensor, and improvements in bimetallic sensors brought the potential for rapid and exact diagnostic procedures to this field, as well as computer-supported evaluation of the results.

Contact temperature readings only register the temperature differences which occur because of regulatory disintegration between reflexly-affected skin areas and non-affected skin areas in one case, and between neutral skin and the point in the other. Comparing these points by measuring only a fixed state gives significantly less information than testing the function, that is, measuring both before and after a cooling stimulus, which allows the type of reaction to be evaluated according to the stimulus.

Since infrared emission (besides a material's infrared radiation to the 4th power of its absolute temperature) is also determined by a skin emission coefficient, infrared diagnosis is more sophisticated, and it also gives more diagnostic information than contact thermometry. Changes in the emission coefficient are mainly caused by the CO_2 content of the capillary bed and by changes in skin texture. It is true that the latter cannot be evaluated as a variation in emission properties, but smoothing a surface brings with it decreased emission per surface-unit of measure while microfolds increase its per surface-unit of measure. While it is true that the skin is generally under autonomic control, this change in skin texture concerns a regulatory system that involves much more than the question of temperature.

Information on the measuring systems and about further studies can be found in the references by Blohmke and Heim for infrared diagnosis, and by Rost for thermodiagnosis.

5.1.2. Electrodiagnosis

In principle, only a few bioelectrical criteria and their changes are registered and evaluated with currently available electrodiagnostic methods:

- the conductance (reciprocal resistance)
- the potential difference between the electrodes, and
- the capacity of the tissue.

Recently fluctuations individual to the body and their changes were ascertained and added to the above list. This is not the place to evaluate or describe the methodology in detail, but the main principles and their physiological basis will be mentioned.

5.1.2.1. The Electro Skin Test

In this test, the skin is stimulated with galvanic power that exceeds the stimulation threshold. The skin in reflexly affected projection zones and at the maximal points turns bright red, and stays red longer than stimulated skin that is not reflexly affected. This method, also called electropalpation, is reported to range from unpleasant to painful, especially on sensitive skin.

5.1.2.2. Conductance Measurement

The first systematic measurements of conductance were, without doubt, taken in the dermatomes levels of diseased organs by Regelsberger, (1952). However, since then, there has been a great increase in the use of point measurement for diagnostic purposes. From a system-dependent standard value, a deviation from a normal value of the conductance upwards is considered a sign of hyperegia (irritation, inflammation, allergy), while a deviation downwards is interpreted as hypoergia (degeneration, regulatory blockade, depletion). The measurements are taken at a variety of acupuncture points—mainly where meridians end in the fingers and toes. In addition, other points are measured in the fingers and toes that are not used in acupuncture. The location of the disturbance or disease is diagnosed according to acupuncture principles and the relationship of the points to organs. *Personal observations show that the musculoskeletal system and its multiple potentials for disturbance are underrated.*

5.1.2.3. Measurement of Difference in Potential

Most practical electrodiagnostic procedures neglect to diagnostically assess differences in potentials, even though it is this change in potential that is usually the primary electrical deviation in functional disturbances and diseases, often leading to secondary changes in the conductance. To date, only two diagnostic methods have included the measurement of differences in potential. One method is used to measure points (Impulse-Dermo-Test), and the other is used to give a bioelectrical diagnostic overview (BF Decoder).

5.1.2.4. Measurement of Capacity

The electrical capacity of the tissue can be measured by applying a defined voltage over a predetermined period of time and by subsequently removing the reverse current by connecting to a shorting circuit to the tissue. However, during the evaluation, the various participating biological components must be considered, since various bioelectrical phenomena are occurring in these experiments, such as uptake of electrons, ion migration, and polarization processes in fixed dipolar molecules

5.1.2.5. Measurement of the Body's Own Electromagnetic Signals

Besides clinically established methods such as EEG, EKG, EMG, etc., since technology has improved in recent years, more and more attempts have been made to record the body's own electromagnetic signals and to evaluate them diagnostically. So far none of these systems has developed enough to recommend it for broad practical implementation, especially since the interpretation of results varies from experimenter to experimenter, and is open to dispute. However, in this connection, one should bear in mind the resonance capacity of organic structures, and that new physiological and pathological knowledge can certainly be expected from investigations in this area (Bergsmann 1983).

6. Towards a Regulatory Therapy

The primary aim of regulatory therapy is to bring disturbed regulatory mechanisms back to their normal functioning, namely to restore the optimal regulatory state and so guarantee the homeostasis and economy of the organism. Treatments that affect regulation can be combined with clinical forms of therapy that conform to etiological thinking, provided that the latter do not produce an additional regulatory stress, such as high doses of corticoids and psychopharmaceutics.

The most important facets of regulatory therapy are:

1. Eliminating factors that disturb regulation, such as minimal chronic stress, heavy metal load, consumption of recreational drugs, excess food, etc.
2. Counteracting possible deficiencies that lead to regulatory dysfunction, such as deficiencies of vitamins, enzymes, trace elements, etc.
3. Breaking down pathogenic feedback mechanisms, for example using acupuncture and/or neural therapy, and also using specific stimulation therapy.
4. It seems entirely possible also to implement regulation-specific therapy using resonance to electromagnetic impulses, but the indications and possibilities have not yet been fully investigated.

Given the scope of this work, a description of all the possible therapeutic methods is impossible. However, special techniques will be informally itemized corresponding to the pathogenetic principles which have been enumerated in this book.

7. References

Athenstaeedt, H. "Pyroelectric and piezoelectric properties of vertebrates." *Ann. N. Y. Acad. Sci.* 238, n. 68 (1974).

Bergsmann, O. "Begünstigen banale extrapulmonale Herden homolateralen Beginn der Lungentuberkulose," *Beitr. Klin. Tbk.* 125 (1963): 506.

_____. *Akupunktur als Problem der Regulationsphysiologie*. Karl F. Haug Verlag: Heidelberg, 1974.

_____. Damböck, E., Glaser, M., Puchas, A., "Der banale extrapulmonale Herd als Gestaltungsfaktor der Lungentuberkulose." *Praxis der Pneumologie* 22, n. 3 (1968).

_____. *Bioelekrische Funktionsdiagnostik*. Karl F. Haug Verlag: Heidelberg, 1979.

_____. "Über muskuläre Resonanz– und Dämpfungsphänomene bei Akupunktur und Lasertherapie." *Dtsch. Z. Akup.* 28, n. 3 (1985).

_____. "Vertebro–respiratorische und vertebro–zirkulatorische Syndrome als leistungsbegrenzende Faktoren bei Degenerationsleiden des Bewegungsapparates." *Rheuma* 7, n. 2.1 (1986).

_____. "Störfeldpathogenese and Relität des Sekundenphänomens." *Rheuma* 7, n. 2.1 (1987).

_____ and Woolley–Hart, A. "Differences in Electrical Skin Conductivity between Acupuncturepoints and adjacent Skin Areas." *Am. J. Acupuncture* 1 (1973): 27.

_____ and Bergsmann, R. *Projektionssymptome—reflektorische Krankheitszeichen*. Universitätsverlag Facultas: Wien, 1988.

_____. "Elektromyographische Verifizierung der positiometrischen Relaxation bei peripheren Spannungssymptomen." *Manuelle Medizin*, 1988.

_____ and Eder, M. *Funktionelle Pathologie und Klinik der Brustwirbelsäule*. G. Fischer Verlag: Stuttgart, 1982.

_____ and Meng, A. *Akunpunktur und Bewegungsapparat—Versuch einer Synthese*. Karl F. Haug Verlag: Heidelberg, 1982.

Bergsmann, O. *Bioelektrische Phänomene und Regulation in der Komplementärmedizin*. Facultas Verlag: Wien, 1944.

Blohmke, M., "Klinische Überprüfung der Thermoregulationsdiagnostik." *Phys. Med. und Rehab.* 7 (1979).

Brückle, G. "Intravitalmikroskopische Utersuchungen über Aufbau und Hämodynamik der normalen terminalen Strombahn der Unterlippenschleimhaut des Menschen." Inaugural dissertation, 1963.

Brügger, A. *Die Erkrankungen des Bewegungsapparates und seines Nervensystems*. G. Fischer Verlag: Stuttgart, 1980.

Drischel, H. *Einführung in die Biokybernetik*. Akademie Verlag: Berlin, 1973.

Gleditsch, J. M. *Reflexzonen und Somatotopien*. WBV Schorndorf, 1983.

Fröhlich, K., quoted in Popp.

Hansen, K., Schliack, H., *Segmentale Innervation, ihre Bedeutung für Klinik und Praxis*. G. Thieme Verlag: Stutgart, 1963.

Head, H., *Sensibiltätsstörungen der Haut bei Viszeralerkrankungen*. Hirschwald Verlag: Berlin, 1898.

Hildebrant, G., "Therapeutische Physiologie, Grundlagen der Kurortbehandlung," in *Balneologie und Medizinische Klimatologie,* edited by W. Amelung, and G. Hildebrant. Springer Verlag: Heidelberg, 1985.

Keidel, W. *Lehrbuch der Physiologie.* G. Thieme Verlag: Stuttgart, 1970.

Kellner, G. "Wundheilung–Mikrowunde (Nadelstich) chirurgischer Laser," in Laser Regulationstherapie. Otsch. Zschr. Akup 22, 86-95 (1979).

Klima, H. Dissertation (Exp. Physik.) Wien, 1981.

_____. "Unbeachtete Informationssysteme des Organismus." Vorgetragen am Symp. d. österr. Ges. f. Neuralth. Baden bei Wien, 1987.

Mackenzie, J. *Krankkheitszeichen und ihre Auslegung.* Kabitzch Verlag: Würzburg, 1911.

Melzack, R., Wall, P.D.. "Pain mechanism: A new theory," *Science,* 150, 971 (1965).

Nordenström, B. E. W. *Biologically Closed Electric Circuits.* Nordic Medical Publications, 1983.

Popp, F. A. *Biophotonen,* revised 2nd ed. Verlag für Medizin Dr. Ewald Fischer: Heidelberg, 1984.

_____. *Neue Horizonte in der Medizin,* 2nd ed. Karl F. Haug Verlag: Heidelberg, 1988.

Regelsberger, H. *Der bedingte Reflex und die vegetative rhythmic des Menschen.* Springer: Wien, 1952.

Ricker, G. *Pathologie als Naturwissenschaft.* Springer Verlag: Berlin, 1924.

Selye, H. *The Evolution of the Stress Concept: Stress and Cardiovascular Disease.* Oxford University Press: New York, 1971.

Travell, J. G. Simons. D. G., *Myofascial pain and dysfunction.* Williams & Wilkins: Baltimore/London, 1983.

Trincher, K.S. *Wasser Grundstruktur des Lebens und Denkens.* Herder: Wien, 1990.

Zwiener, U. *Patholophysiologie Neurovegetativer Regelungen und Rhythmen.* G. Fischer Verlag: Jena, 1976.

Part Three

1. The Therapeutic Consequences of Matrix Regulation Research

1.1. The Puncture Phenomenon

It was always important for Pischinger to point out that even minor injuries—such as the prick of a needle—whether for drawing blood, acupuncture or for neural therapy—lead to a clear reaction in the extracellular matrix.

Usually in the hours after blood has been drawn there are large changes in iodine consumption values.[4] As Pischinger previously stated, the loss of 3–5 ml of blood cannot be responsible for this effect. Only the *puncture itself* can be responsible—the puncture that affects the perivascular tissue, which, as is well known, is a loose connective tissue richly supplied with autonomic nerves. This is why I have mentioned a puncture phenomenon or a puncture effect. This puncture makes a wound, if only a small one (Kellner 1979). Besides this, there is even a change in potential from the needle to the tissue—as will be discussed—which is in a range that can be measured.

This puncture effect also manifests in other characteristics of autonomic function, and is an essential quality in the nonspecific arena. This will be discussed in greater detail and supported by experimental results.

Here are some examples of the occurrences of this effect during iodometry. In order to achieve our goal, we must become as familiar as possible with what happens during venipuncture. A needle *puncture* travels (1) through the epithelium of the skin, (2) through the loose connective tissue of the stratum pappilare, (3) through the firm tissue of the stratum reticulare, which has connective tissue interspersed along nerves and blood vessels, (4) through the perivascular tissue of the veins which also has loose connective tissue, (5) the muscular wall of the vein, and finally, (6) the intima and (7) the blood. So there are plenty of locations available from which an autonomic reflex can be stimulated. Reactions particularly originate in the perivascular system, which contains the basic regulatory triad—including extracellular fluids (in regards to this, refer to G. Kellner's skin experiments.)

In the case of syringes or acupuncture needles, a puncture affects the body in 3 ways: (1) the smallest injury has the longest-lasting result (at least 5 days according to Kellner and Feucht), (2) through the difference in temperature between needle and tissue (Needle: room temperature, ca. 20–22°C, tissue: between 36 and 37°C.) and (3) the difference in potential between needle and tissue. According to Bethe, Gildemeister, Hauswirth, Karcmar, Neuberger, Croon, Overhof, and Maresch, this can be measured. Electrical stimuli spread very rapidly over the matrix system to the entire organism. Thus, a photon emission from

---

[4]. Iodometry is used in these studies to quantitatively measure oxidants (such as free radicals) as they arise, e.g., from phagocytosis or from stress. Iodine consumption value = ICV.

the injured, or to be more precise, the dying cell, reaches the neighboring cell in 10^{-7} second, and spreads over the entire organism at least at the speed of sound (Popp 1984).

Kellner, using an infrared camera, and Bergsmann, using rheography, clearly verified the effect of the temperature difference.

Figures 1 and 2 show a series of puncture reactions in graph form (see Part One), which show a good deal of variation. When compared to a zero point, they present an uncommonly variable reaction pattern, which is caused by the various immune states that the individual subjects are in. These results can then be used in testing the immune state.

The reaction of a healthy subject corresponds to the pattern of the alarm reaction. According to Selye, a stimulus of this amount will be completely smoothed out through regulation within 3–4 hours. In the first hour, the iodine consumption value in the iodometry sinks, according to Pischinger, so that in the following 2–3 hours, it can again rise to the final value, in what may be called a shock- and countershock reaction.

A disturbed reactivity of the non-specific system leads to other reaction forms (RF) which are also subject to certain principles: extremely high iodine consumption values (ICV) appear in conditions such as sudden allergies. In this case, the puncture reaction can trigger an asthma attack. A decrease in the ICV can be observed in conditions such as delayed allergies (i.e., inflammatory rheumatoid conditions, multiple sclerosis, ulcerative colitis, etc.). Reactions are completely absent during all-consuming illnesses (e.g., malignancies, leukemias, results of thorotrast exposure), as well as in late and end stages of inflammatory illnesses (such as tuberculosis or chronic progressive systemic illnesses).

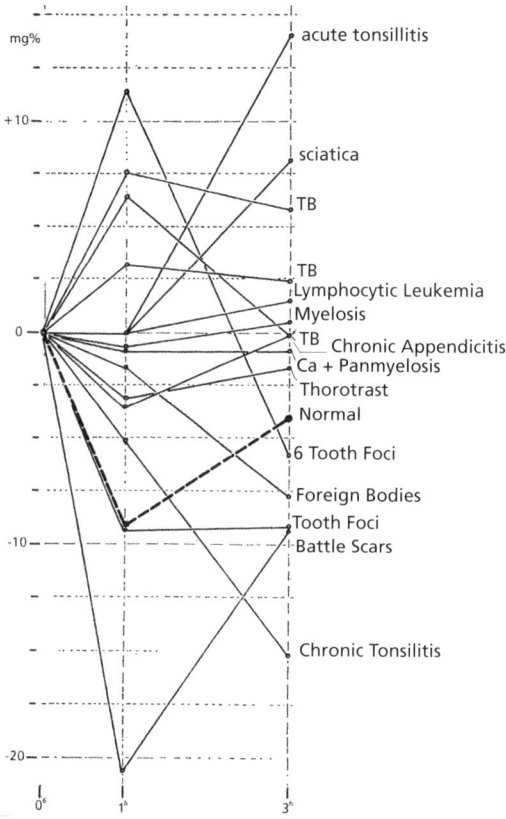

Figure 1. Reaction variations during three-hour test of stress during the drawing of 5 ml blood from the cubital vein. All beginning values are recorded as "0" the deviations in the following hours as "+" and "-" values in mg% iodine consumption. Abscissa: time, in hours, after drawing blood.

It is clear from this experimental group that the non-specific system reacts in a highly sensitive manner. The puncturing itself triggers changes in amplitudes, in this case in the ICV. Skeptics occasionally asked why the puncture of the second blood drawing didn't trigger a reaction analogous to that of the first. It should not be ignored that—as demonstrated by the all-encompassing experiments of Lickint as well as Selye—an organism that is in the process of *an alarm reaction* does not react to further autonomic influences.

Experiments show that, when coupled with standard mild stresses such as puncture, the characteristics of a protein-free or protein-poor serum extract, if gently obtained, can be used to test an organism's *reactive state* and immune capability. Obviously this is a humoral reaction which, however, could not occur without the other specific areas of the autonomic system. It should be mentioned in passing, that this is a manifestation of the basic phenomenon of acupuncture.

It seems to be most important to ascertain that even a minor irritation to the affected tissue, that is, a puncture, sets the entire reaction system—this means primarily the non-specific matrix regulation system—in motion. Pischinger speaks of the *puncture phenomenon that manifests the holistic character of the matrix system.*

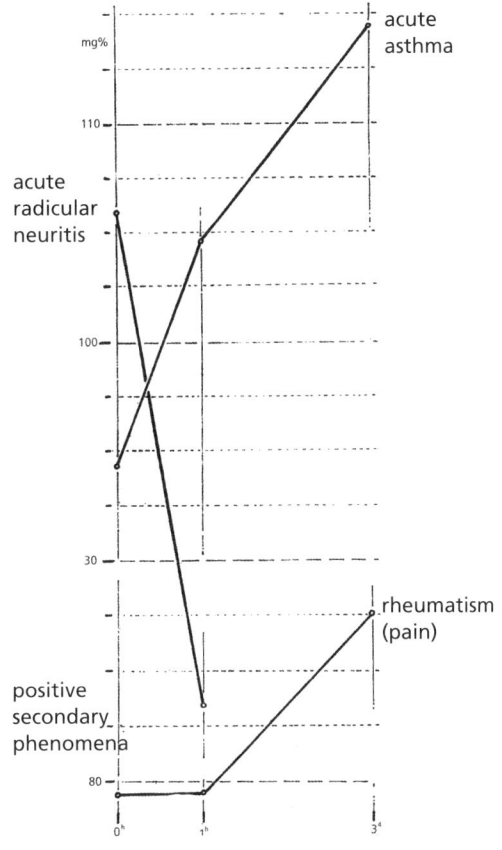

Figure 2. Puncture effects. With the increase in the iodine consumption values, acute attacks of asthma and sciatica are set off. In acute radicular neuritis, permeating the scar with Impletal caused a decrease in the previously excessive iodine consumption values and also decreased pain.

Table 1

	Time	Erthro.	Leuko	Segm.	Eo.	Baso.	Mono.	Lympho.	BF
I	8	5,320,000	8366	5020	250	0	500	2430	160
	9	5,370,000	8832	4240	179	88	353	3000	972
	11	4,800,000	6978	3980	69	69	628	2023	209
II	8	4,400,000	4100	2255	41	41	164	1435	164
	9	4,330,000	4000	2160	40	40	160	1400	200
	11	4,200,000	4100	2091	41	41	164	1435	328
III	8	–	5100	2300	50	50	150	2000	550
	9	–	4466	1700	45	90	135	1956	540
	11	–	6400	2956	0	64	320	2100	960

BF = breakdown forms

Table 2

Skin	R	C	
Undamaged	1.30	0.069	
1 puncture point	1.75	0.07973	R in kiloohms
3 puncture points	2.11	0.145	C in microfarad
7 puncture points	1.76	0.380	
14 puncture points	1.33	0.940	

These deviations from the normal reactions also demonstrate major differences in the magnitude of the breakdown forms of leukocytes (leukolysis). In a healthy subject, a simple puncture stimulus causes more than a five-fold increase in lysis forms (Table. 1, Pischinger 1983). The humoral-autonomic blockade (*Blockierung*) which is shown in case II (Table. 2), only shows an approximately 22% increase in lysis forms in the first hour, and a 100% increase after 3 hours—a greatly retarded and meager reaction. Case III has an even more sluggish and restricted reaction. In the first hour there is no increase in lysis forms, and after three hours, a slight increase of just under 70% (Table. 3).

Table 3

Patient	Date of measurement	Time of measurement	Initial state		Puncture reaction	
			R	C	R	C
Dita Sch.	3.1.69	18:35	11.1	0.15	—	—
		12:20	—	—	16.7	0.21
		12:23	—	—	16.3	0.23
		8:35	11.5	0.16	—	—
		12:20	—	—	17.9	0.14
		12:23	—	—	17.2	0.14
		8:35	6.1	0.23	—	—
		12:53	—	—	9.7	0.20
		12:55	—	—	8.8	0.23
		8:10	8.5	0.21	—	—
		12:25	—	—	14.0	0.18
		12:27	—	—	12.3	0.18
Josef Tr.		8:14	6.3	0.23	—	—
		12:57	—	—	9.5	0.25
		12:59	—	—	6.7	0.29
		8:49	5.4	0.27	—	—
		12:57	—	—	3.2	0.31
		12:59	—	—	2.9	0.44
Erna D.		8:35	16.2	0.15	—	—
		12:47	—	—	23.1	0.16
		12:49	—	—	15.3	0.19
		8:43	13.3	0.12	—	—
		12:52	—	—	13.7	0.21
		12:54	—	—	10.1	0.23
Peter J.		8:30	13.3	0.12	—	—
		12:20	—	—	24.1	0.10
		12:22	—	—	16.1	0.14
Sr. Gebharda L.		8:50	9.0	0.21	—	—
		12:29	—	—	13.0	0.18
		12:31	—	—	14.8	0.13

R in kiloohms
C in microfarad

The meaning of these variations was unclear. Later experiments that simultaneously determined immunoglobulin A, M, and G, showed three details that clarify the meaning of leukocytolysis:

- The changes in the immunoglobulins (especially in IgG) depend on the degree of leukolysis. Minor leukolysis signifies nothing besides minor rate of leukolysis, but high lycolysis is accompanied by a major increase in the immunoglobulins (especially of IgG) which indicates that plasma cells are breaking down.
- The breakdown of monocytes which releases, e.g., increased amounts of triple-conjugated unsaturated fatty acids (Factor M according to Pischinger), which are partly responsible for the triggering of humoral shock conditions and take part in the changeover to the counter shock reaction (the phase of activation of the immune system);
- The breakdown of granulocytes, which—as in the microphage phase of the local immunity—releases oxydative and proteolytic enzymes as well as interleukins, prostaglandins, and leukotrines, and thereby neutralizes living irritants (Perger 1990).

In the shock phase of the puncture reaction, the ICV falls approximately 10%, but at least down to 50 µg/ml. However, it rises back to the initial value in the following 3–4 hours.

However, the reactions to simply a puncture are demonstrated not only in iodometry and in leukocytolysis, but also in other criteria of the non-specific system, such as in oxymetry of venous blood, in fluctuations of minerals in situ and in venous blood, and are even found in the immunoglobulins, when these also proceed sluggishly and at prolonged pace.

The puncture, however, also triggers bioelectric changes.

1.1.1. Bioelectric Events During Puncture Phenomenon

The bioelectric phenomena during the puncture reaction indicate processes which are important for the orientation of the immune system (Kellner 1979).

As Kracmar (1971) reported, Diehl (1937) had previously described changes in the polarity of the skin when it was punctured with a needle.

Gildmeister (1928) and Lullies had already demonstrated that the polarization qualities of the skin can be understood through a Wheatstone bridge supplied with alternating current.

Diehl (1937) used a frequency of 756 hertz. Kracmar (1971) later used 50 hertz, since the polarization phenomenon becomes more evident with this frequency, according to Gerstner. Hand to hand resistance (R) is measured in kiloohms and capacity (C) in microfaradays (Kracmar 1971).

Here too, the reaction form depends on the immune state the individual was in at the time of the investigation. Kracmar established that the same reactions occurred with acupuncture needling.

These bioelectrical phenomena are accompanied by thermoregulatory changes. After needle punctures, the skin temperature changes. Kellner (1971) confirmed

The Puncture Phenomenon ✦ 1.1

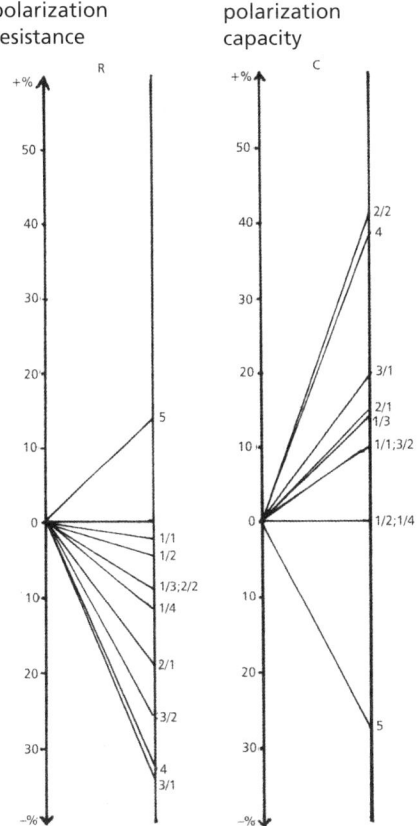

Figure 3. Changes in the polarization resistance and the polarization capacity to puncture reactions (Prof. Kracmar). Left: Pole resistance. Right: Pole Capacity.

Initial values are always start at 0. Variations are recorded as "+" and "-" values.

Figure 3 and Table 3 show puncture reactions in various patients before the beginning of the therapy, which Plohberger and Kracmar (1970) carried out during experiments which employed the cancer therapy of Leupold.

this in acupuncture with an infrared camera. Using rheography, Bergsmann (1965b) was also able to demonstrate the change in blood flow in the skin.

These bioelectrical and thermoregulatory phenomena, which take place after the skin is punctured, are important in describing the functions of the matrix system. They indicate that biological reactions work on a basis of electronic-energetic processes, and that life itself is not possible without these prerequisites. Focusing on morphological changes and purely biochemical processes has certainly brought a great deal of valuable knowledge, but it fails to explain life itself, and what is clinically more important, fails to explain questions of energy and its activation in all living processes, including the immune functions.

Recalling that each exogenous stimulus not only triggers cellular and humoral reactions, but is always accompanied by alterations in the biopotential, one has sufficient cause to focus much more on these biopotentials than was done in the past. Of course, it is not necessary to become obsessed with the point that these biophysical processes are absolutely necessary energetically for the diverse functional processes—but it is becoming more and more difficult to suppress this opinion and refute it experimentally in the course of cybernetic investigations.

However, be forewarned of a great temptation: over and over again one sees that, based on physical examinations such as these, conclusions are drawn that give a direct specific diagnosis, as in the case of silent inflammation and similar conditions. This, however, is inaccurate. While it is true that decoder dermography and thermoregulatory diagnosis can indicate where a disturbance is, they cannot replace subsequent specific diagnostic methods which determine what type of disturbance is involved. They can only document the state of the nonspecific system.

The entire significance and problems involved in the different bioelectrical processes are presented in a separate contribution to this book by Bergsmann. His point of view is that the puncture phenomenon is not only crucial in acupuncture, but also influences neural therapy and every type of injection therapy—this opinion is also supported by the results of humoral investigations (e.g., the difference between injecting factor M and a transdermal application). In addition, it should also be pointed out that as predominantly humoral medical practitioners, in order to learn to understand the processes involved in chronic illnesses at all, one must while investigating matrix regulation unhesitatingly confront the reactions bordering between biophysics and biochemistry—which must become expanded through biophysical results. Similarly, it is obvious in humoral research that those reactions that occur in the area bordering between biophysics and biochemistry—e.g., oxygen consumption, the reactions in factor M according to Pischinger, in the electrolytes, and so on—proceed more rapidly and more sensitively than biochemical processes that are only controlled by enzymes. These are facts that must be recognized.

1.1.2. Puncture Phenomenon and Oxygen Saturation of the Blood

Pischinger (1974) called the changeability of the oxygen saturation of venous blood as the most important sign of the manifestation of the "holistic reaction" in the matrix functions.

Of note is that when venous blood is drawn, it can often vary between being a very dark red and an arterial bright red color. The high degree of arterial coloration is often noticeable during rheumatic inflammatory flare-ups and multiple sclerosis exacerbations. This is certainly not always the case, but it is so common that this observation can be attributed to a disturbance of oxygen consumption, and undoubtedly has significance in the course of an illness. Pischinger (1954) began to follow this phenomenon using an AO oximeter (*Hellige*). Much later (between 1980–1986) we were able to continue in this manner using our own criteria, and indeed confirmed his findings.

The puncture itself leads to the same reactions which are fundamentally the same in oxygen consumption, i.e., in the oxyhemoglobin content of venous blood, iodometry, the leukolysis reaction, and the R and C measurements (resistance and capacity), etc. This once again proves that the response to a stimulus depends on the state of the individual's immune system at the time of the stimulation.

Repeated testing of the oxyhemoglobin content of venous blood in healthy subjects showed first a resting level (at least 20 minutes before taking blood) of about 40% (35%–45%). The exogenous stimulus leads initially to an increase in venous blood oxyhemoglobin as a sign of a humoral-autonomic shock reaction. After 3 hours, the stimulus is again balanced out with a return to the initial value. In chronically ill patients, the initial values are completely different. What is first noticeable in the exudative stages of inflammatory systemic diseases are the considerably increased oxyhemoglobin values, which on the average are 75%, but occasionally go as high as 92%. Considering that the oxyhemoglobin in the arterial part of the vascular system is between 96% and 98%, such values indicate that oxygen delivery to the peripheral tissues has been significantly reduced. The cause for this could be a massive opening in arterio-venous anastomoses, which Bergsmann (1965a) demonstrated in his research on blood perfusion in diseased organs. The puncture reaction briefly reduces the blood perfusion even more. The reverse, an increased consumption of oxygen, is found in chronic progressive inflammations. Here, the oxyhemoglobin levels go below the normal level of 40%; in extreme cases they go as low as 3% (!). Even this can be explained by the previously mentioned work of Bergsmann. This extreme reversal of oxygen consumption in the periphery can only be explained theoretically: in exudative disease processes, a higher oxygen content would probably be associated with massive tissue destruction, and therefore it would be suppressed. In the chronic-progressive stage, however, energy is lacking, and there is an attempt to compensate for this by increased delivery and use of oxygen. This assumption is based on the fact that the human organism is an open energy system (Heine 1987), which is subject to a biological balance of flow with highly interlinked regulatory systems (v. Bertalanffy 1952). Every organism is constantly attempting to achieve this balance of flow, and in addition, trying to get the energy it requires—so there is some justification for regarding the increased oxygen consumption as a substitute regulatory cycle to obtain energy.

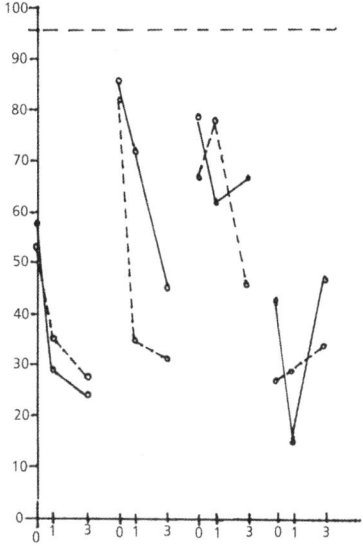

Figure 4. Bilateral determination of O_2 content. Solid line is the right cubital vein; dotted line is the left cubital vein. 3-hour individual tests. Asymmetrical reactions.

1.1.3. Iodometry and the Puncture Phenomenon

The results of the iodometry, according to Pischinger, which have already been presented extensively in this book, show that puncturing a vein produces the same type of response as a shock reaction. In healthy subjects, as has already been mentioned, the ICV decreases by at least 50 µg/ml. This decrease is followed in the first hour by a reverse regulation which restores the initial value within 3–4 hours; this has also been mentioned previously. The reactions of patients with regulatory disturbances are varied in degree and kind. Not only do subsequent blood values greatly exceed or fall short of the normal initial value of 810 µg ICV, but fluctuations (increases and decreases) differ greatly according to the form the disease process takes. The duration of the reaction to the puncture also shows just as much variation.

1.1.4. Totality of Regulation During the Puncture Phenomenon

As Pischinger (1975) already described in detail, the puncture stimulus affects the following series of parameters of the humoral-autonomic arena: (1) the differential blood count, (2) the leukocyte count, (3) iodometry, (4) oxygen utilization in the periphery, and (5), the electrical parameters in the skin. In all these parameters, there are distinct reactions in the sense of a shock phase of the alarm reaction. These reactions, however, depend on the what the specific individual state was initially. As parameter 6, use of capillary microscopy, Pischinger brings up the subject of functioning of the matrix system (Brückle 1969b).

The various types of reactions which arise from the differing initial states also indicate that this highly interlinked open energy system called a "human being" is mainly concerned with maintaining life, and with preventing the potentially fatal effects of a direct linear, cause-and-effect series of events. It is therefore impossible to establish such linear causalities, particularly in chronic diseases. One must start with this premise in order to make it possible to even have a hope of understanding of the various types of responses to stimuli, or of being able to treat them in an analytical manner.

Every stimulus that exceeds local immunity triggers a reaction in the *entire* regulatory system, that is, in the entire intercellular-extracellular matrix. It is intimately connected to all the other regulatory systems due to its very close relationships (in other words, interlinking) via the capillaries, the lymphatics, and the autonomic nerve fibers. These conclusions lead to a *new* concept of the relatively holistic functions in the organism, where the "whole" is greater than the sum of its parts.

The stimulation threshold for these holistic reactions is relatively low, as the puncture phenomenon demonstrates. As will be shown below, these thresholds can vary a great deal in illness and in health. It is true that the reactions are holistic, but—and at first this was a surprise—they are not the same throughout the organism.

2. Testing the Initial State and Autonomic Asymmetry

Recording the immune reactions during chronic diseases, particularly in inflammatory systemic illnesses of unknown origin, was not only theoretically interesting, it was advantageous in treatment.

At first (Perger since 1949), using long-term hemograms [CBCs and blood differentials], attempts were made to understand the process of the active episodes of illness, and over a longer period of time, to understand the chronic, progressive reaction processes of multiple sclerosis (MS). Certain principles were actually found. However, at that time the research conditions were very limited (e.g., serum electrophoresis was not introduced until 1952). Essentially, the only measurements that were at first possible were the following: evaluation of the hemogram and, from this, determination of the absolute values for eosinophils; calculation of the electrolyte values for eosinophils; measurement of the electrolyte value (Ca, Mg, later K); and total cholesterol. It was known that calcium and eosinophils play a role in allergies and allergic-type processes and, in addition, a paper by Aiginger was available on the variations in the magnesium level in the active and quiescent stages of MS. However, even these limited means made it possible to establish the most important types of reaction in the field of nonspecific immunity. The following graphs show the various reactions during the process of both episodic and chronic-progressive MS, as well as their very different reactions in acute febrile infections; the latter in an influenza epidemic in 1954.

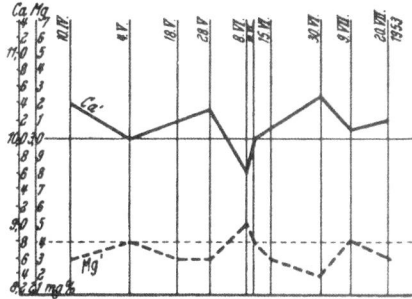

Figure 5a: Ca-Mg curve in a case of MS in remission (normergic reaction). Calcium: ——, Magnesium: - - - - -. (thin horizontal lines: for Ca 10.0 mg%, for Mg 2.4 mg% indicate the critical level of the ion values (*see* text).

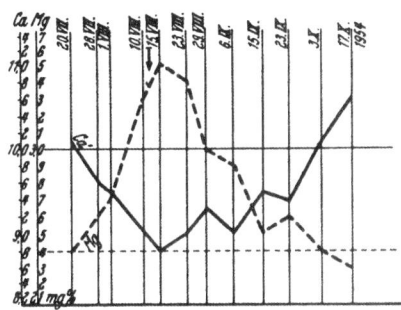

Figure 5b: An especially typical Ca-Mg curve before and during an acute MS episode. Since July 26, 1954 subjective complaints of malaise and insomnia. First clinical symptoms on August 13, 1954. On August 23, 1954 symptoms stable with subsequent remission.

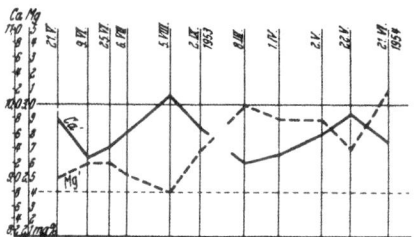

Figure 5c: Long-term progression of Ca-Mg levels in a case of a chronic-progressive form of MS, observed for 1 year. Ca remains less than 10.0 mg% and Mg above 2.4%. (Figure 5 a-c, from Perger, DMW 81 [1956] 342 (Allergy Supplement])

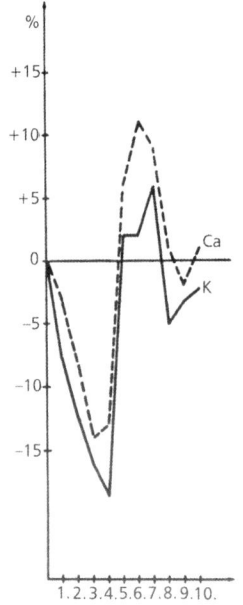

Figure 6: Difference in the electrolyte level (mg%) as a percentage of the initial value during acute infection (from: F. Perger, Vienna, med. Wschr 12 [1978], 31–37

Initially, three of the possible types of reaction were recorded. In the process, the normal, strong reaction of a healthy immune system during acute infections also suggested a potential interpretation for the results. It showed rapid and ample fluctuations in the electrolytes with some minor after-fluctuations during convalescence. This corresponded to the well-known alarm reaction that had just recently been described by Selye (1953), with shock, countershock, and convalescence phases. The whole process lasted between 7–10 days, and the total range of fluctuation was about 40% in the transition from the shock to the antishock phase.

During acute MS episodes, there was no sign at all of this reactive triad. The entire clinical process was accompanied by an electrolyte pattern that corresponded to the shock phase during acute inflammation. Only occasionally in early cases was there any sign of a rudimentary antishock phase. Generally, an MS episode goes to completion displaying a humoral pattern of a pure shock phase. However, compared to those in acute inflammations, the criteria fluctuations in acute MS were correspondingly somewhat less (maximum of 25% in the electrolytes) and the duration was significantly longer (by at least 40–50 days, i.e., 6–7 weeks). And in the chronic-progressive form of MS, no basic pattern change at all was observed in the criteria, particularly in the electrolytes—they stayed mostly within the reaction levels that are found in the shock phase of the acute and exudative episodic type of reaction.

In this way, the three most important forms of reaction (RFs) were found, which are still as important today as they were then (*see* Fig. 8): the normal RF (reaction

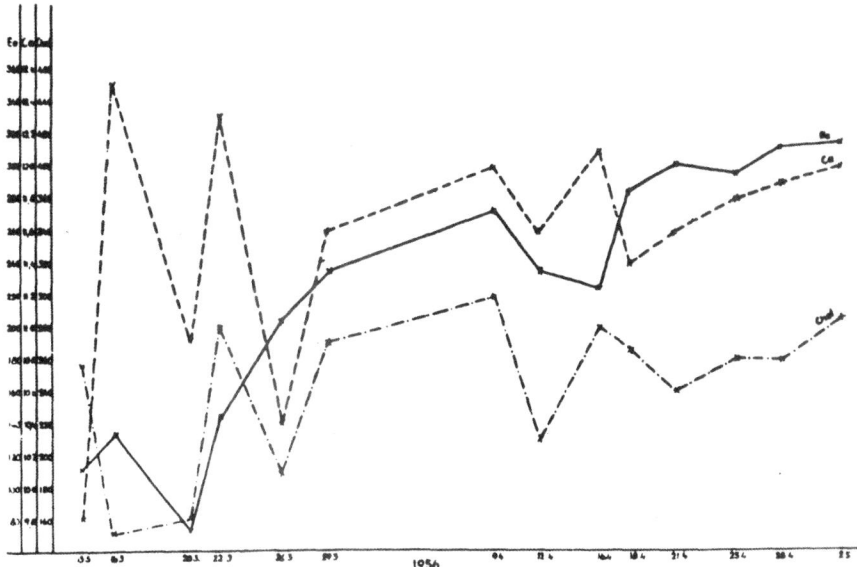

Figure 7: The changeover from "ataxic" (uncoordinated) to chronic reaction types in a case of untreated polyarthritis in a 20-year-old woman (from F. Perger, Ther. Wo. 8 [1958], 224)

form), the RF of an exudative exacerbation, and the RF of chronic-progressive inflammations (1990). From 1954 on, the same long-term blood counts used in MS and acute febrile infections were also applied to internal chronic diseases. In the process, individual transitional forms from normal to pathological RF were found.

In particular, a fourth important RF was found, in which an episodic, proliferative-degenerative RF transforms into an isolated antishock phase. The course of these episodes lasts a very long time. The longest episode that was followed up humorally lasted 231 days (i.e., 33 weeks). This RF comes very close to blocking the matrix regulation.

Serum electrophoresis confirmed the findings. The progression of changes in globulins is completely connected to the nonspecific RF. The development, particularly of the increase in γ-globulins (IgA, IgM, and IgG) depends on the process of nonspecific regulations. This increase is always at its maximum at the end of the inflammatory nonspecific reactions, i.e., after 7–8 days in acute inflammations, after 6–8 weeks in episodes that have an exudative character, and only after 25–35 weeks in proliferative-degenerative reactions.

This was the first indication that the speed and intensity of the specific immune reaction is intimately related to matrix regulations. It is further evidence that matrix regulation is crucial and that it obviously presents or, to be specific, *embodies* a system of order. In favor of this is the fact that pathological immunoglobins—for

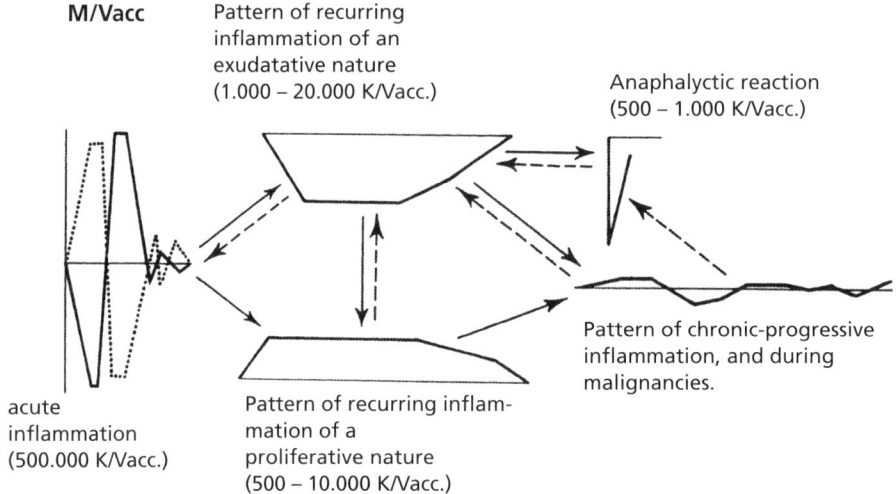

Figure 8: Diagram of nonspecific reaction types according to Perger, corresponding to the similar calcium and cholesterol reactions, and including potassium reactions during regulatory limitations. The figures in parentheses indicate just the quantity of vaccine (M: microorganisms) needed to barely initiate the reactions presented here (from F. Perger, Phys. Med. and Rehab. 20 [1979], 585).

example, from serum samples in rheumatoid arthritis—indeed are, on occasion, temporarily positive in the episodic forms, but they are only long-lasting when there is blockade of the matrix functions and the course of the illness is a chronic-progressive one.

The absence of a reaction in the criteria of the nonspecific system was already called "blockade" by Lutz and Pischinger (1949), although the term "paralysis" was closer to the mark. However, there is a clear difference between these two terms. Rigidity of nonspecific regulation can be temporary, and can also resolve spontaneously (e.g., after short-term chemotherapy), or can be resolved therapeutically (e.g., by neural therapy and acupuncture), but there are also regulatory rigidities that cannot be resolved (e.g., in primary chronic [rheumatoid] arthritis), and there is justification for calling these regulatory paralysis. Equally well, the regulatory rigidity in malignancies and the end-stages of chronic inflammations such as tuberculosis is true paralysis of function.

Regulatory blockades do not present a consistent pattern like the other types of reaction, and an attempt must be made to differentiate between resolvable and unresolvable blockades of the matrix system. This can often be accomplished simply from the specific clinical picture, but it can be objectively determined by the ICV level in iodometry. Even though it can be difficult to do, greatly elevated ICVs (ICV between 900 and 1,000 µg/ml) with blockade can usually be resolved. If the ICV values are lower, however, they cannot be resolved (ICV <780 µg/ml). It is therefore important to assess the extent to which these "blockades" can be resolved

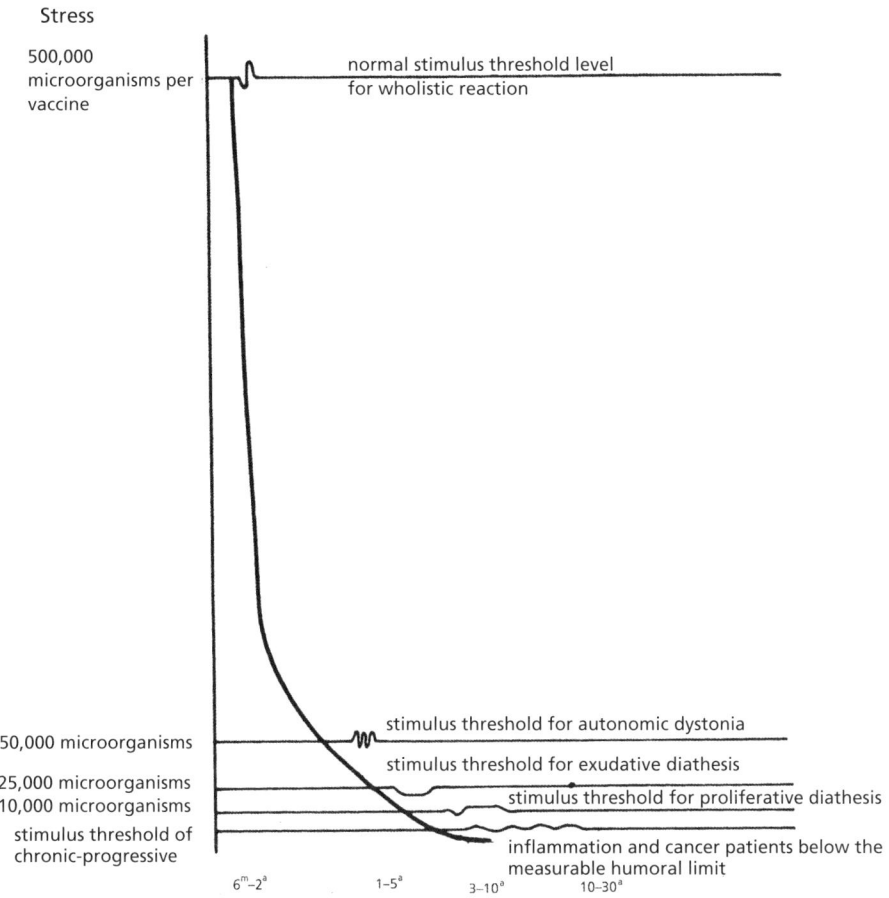

Figure 9: Provisional diagram of stimulus thresholds for holistic reactions in healthy subjects and in patients with chronic, recurring and chronic, progressive diseases (from F. Perger, Entretiens de Monaco 1980, "Le role de la médecine moderne dans las crise du monde occidental," p. 49–59, Clubs médecine informatique, Denisé, France, 1981).

because this determines the theory that treatment will be based on—rehabilitation of the immune functions or immunosuppresive therapy.

After sorting out these process types in 4,716 patients and after a minimum of 8 years of follow-up observations (Perger 1978), the next logical step was to determine how the patients react from the very beginning. In addition, it was necessary to develop an effective short-term test.

Initially, this was tried with vaccines of various types of microorganisms, including autovaccines and heterovaccines. However, the results were completely different than what was intended. Using the vaccine test, instead of being able to establish the types of reaction, what was found were the stimulus thresholds for holistic reactions. True, these thresholds depended on the types of reaction in the long-term hemogram, but they gave no information about the existing RFs.

In total, the 917 vaccine trials showed a pathophysiological phenomenon. They showed that the periphery was increasingly unable to regulate and locally override stimuli up to a certain level. In healthy subjects, the stimulus threshold is relatively high, approximately 500,000 microorganisms per vaccine. However, in regulatory disturbances, the threshold sinks rapidly, and finally drops so low that it can no longer be measured by humoral testing methods. This means that in chronic diseases, the entire immunity of an organism must be activated sooner and faster—a completely uneconomical use of energy, which in time leads to its overuse (Bergsmann 1965b).

The results of these vaccine stress trials does, however, have clinical significance, because, for example, in desensitization therapy, the reductions of the stimulus threshold are taken into account, avoiding possible overreactions. However, the aim of obtaining a quick method of establishing the reaction type remained elusive.

The solution to this problem came from a clinical trial on how the monocyte factor and the triple-conjugated unsaturated fatty acids worked, and this expanded the available research that already had already been done by Pischinger (Perger 1956). As described in previous sections, raising the amount of monocyte factor in the blood by injection produces a significant increase in the number of monocytes. However, this means there is a changeover from the microphage phase to the macrophage phase of cellular immunity. This also signifies a simultaneous change from the humoral shock phase to the countershock phase, which in acute infections means a simultaneous change from the prodromal phase to the active immunological phase. Humoral shock is resolved by the physiological increase and the distribution of these particular fatty acids. Because of this special ability, an injection of this type of fatty acid seemed to be a way to bypass the primary shock phase and begin the active immune processes of the countershock phase. This idea was successful.

After the first time blood was drawn, factor M (Elpimed®) was immediately injected subcutaneously and the changes of the criteria were measured after 1 and then 4 hours (afterwards, they were measured after 3 hours) (Perger 1963a).

It was already known from the long-term hemograms that all regulatory disturbances can be diverted by Selye's alarm reaction. These deviations have already been discussed, but will be mentioned again as follows: being stuck in the shock phase (exudative-allergic-type process forms), loss of the shock reaction and lapse into the countershock phase (proliferative-degenerative process forms), and regulatory rigidity (blockade, i.e., chronic-progressive process, malignancies). In addition, there are transitional forms from normal to pathological RF, and the anaphylactic reaction, which is an acute, severe reaction within a few minutes.

Injection of factor M actually bypasses the shock phase which, however, cannot be avoided if foreign substances have been injected. And within 3–4 hours, an

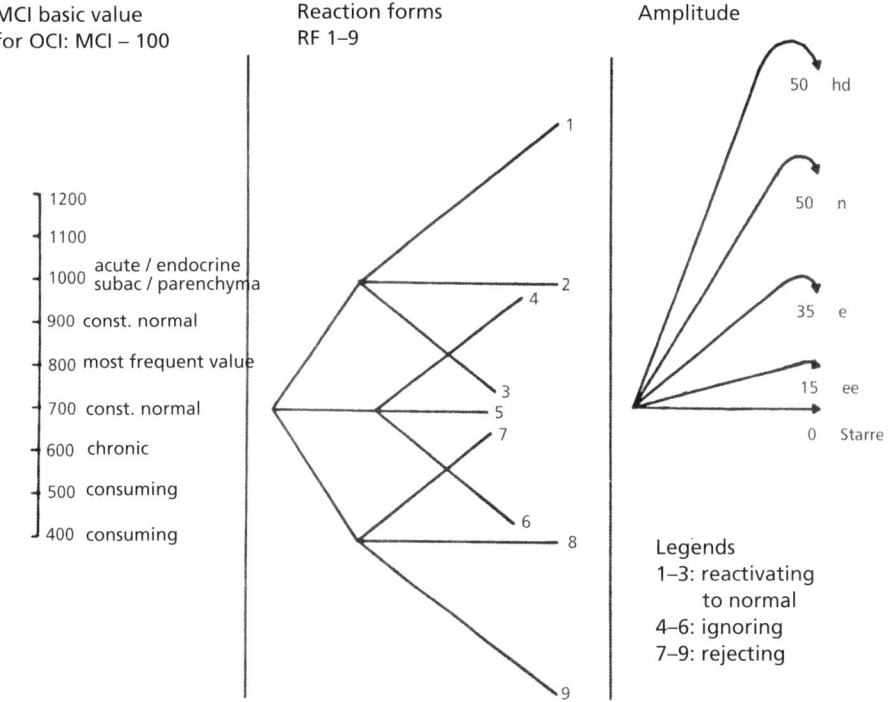

Figure 10: From Kellner, Krammer, Seidl, Die Heilkunst 91 (1978).

organism shows its ability to deal with an immune-stimulating stimulus.

There are a variety of reaction forms (RF), which Kellner (1979) has systematized on the basis of a statistical evaluation of 1,200 such trials.

Once again, a healthy organism can regulate and recover from a moderate stimulus within 3–4 hours (RF 3). Simple blockades of the matrix functions, e.g., after a common cold, after chemotherapy, and in the initial fleeting manifestations of an inflammatory systemic disease, lead to stronger and longer-lasting ICV reactions which are a sign of greater activation of the immune system. In delayed recovery from a cold, however, one can sometimes trigger another febrile event, which resolves the remaining infection. However, any improvements in functional disturbances and manifestations of systemic inflammation are only temporary, since factor M cannot control the causes of theses processes (silent, chronic inflammations, scar disturbance fields, subclinical toxicosis, etc.). Care is recommended in allergic diseases such as bronchial asthma. It is certainly unusual, but both Pischinger and the author of this chapter triggered attacks of bronchial asthma with this injection—in both of my own two cases, the cause was a bacterial allergy. These reactions include RF 1 and

2, with excessive and long-lasting elevation of the ICV.

However, already-manifested diseases of the delayed allergy type (inflammatory forms of rheumatism of the joints, MS, ulcerative colitis, etc.) react after one hour, with intensification of the humoral shock phase (RF7–9). This contradicted the finding that factor M is the body's own antishock substance. There was a solution to this problem: Drawing blood as soon as 30 minutes after the injection established that these special fatty acids had the effect of triggering shock (Fig. 11).

It took these 30 minutes for the shock reaction to appear. This confirms the concept of the open, highly interlinked system described in the previous chapter. When the noxious factors that set off the disease cannot be overcome, the activation of immunological functions is *actively* obstructed. This occurs in the efforts that are central to an open energy system such as the human organism, i.e., to maintain life and avoid major damage. If an organism reckons that activating the immune processes is dangerous, it will use every opportunity to undermine such activation. These processes involve a substantial expenditure of energy. In any case, this detail must be seriously considered in chronic disease.

This also makes it understandable why at times there is no reaction at all after a stimulus, in fact, in any of the parameters of the immune system. Regulatory blockades, which are RF –6 in Kellner's diagram, not only affect the ICV; they also affect the electrolyte levels, fluctuation in the lipids, etc. Here, it must be emphasized that, e.g., the electrolyte levels, indicate blockades

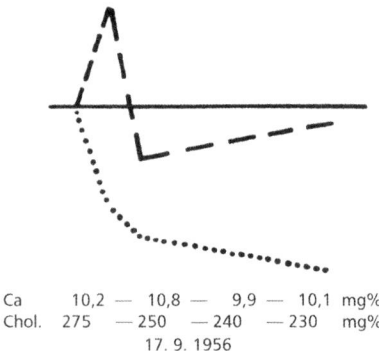

Ca 10,2 — 10,8 — 9,9 — 10,1 mg%
Chol. 275 — 250 — 240 — 230 mg%
17. 9. 1956

Figure 11: Accelerated reaction process: Even if blood is drawn as soon as 30 minutes after a stimulus, the primary antishock can be measured (from F. Perger, Öst. Z. f. Stomat. 60 [1963] 440).

much earlier than do the ICV or oxymetry. All these are signs of energy depletion in the matrix system, but not total depletion; reactions which are too severe continue to be blocked as much as possible. What the chronic progressive course of inflammatory diseases shows is that this blocking is no longer fully possible. Complete energy depletion is only seen shortly before death, as in the terminal stage of cavernous tuberculosis, which showed a normalization of all parameters a few days before death. In any case, the depletion is so great that the minimal stimulus of a test substance produces no reaction.

These trials were finally published (Perger 1963), based on 435 trials using only Elpimed®, and a further 69 using Elpimed® in combination with other medications. This had scarcely been published when Bergsmann (1965) reported on the asymmetry of autonomically controlled factors. He established that there was a significant difference between the leukocyte counts in the two halves of the body first in unilateral tuberculous processes, and later also in unilateral focal stress. He assumed that the asymmetry was due to a difference between the two sides in the blood flow through the vascular bed, possibly because of the functional condition of the arteriovenous anastomoses. This statement was and still is improbable. The counter-argument, which continues to this day whenever physicians are confronted with this problem for the first time, is that the heart continuously mixes the blood, so different cell counts cannot possibly exist. However, this argument only works for the blood in the arterial part of the vascular system, and not for the blood that has gone through the peripheral capillaries into the veins. The tissue in different locations detects in entirely different ways what the arterial blood has to offer, and the blood releases what it can and indeed must release.

Pischinger, along with Kellner, not only confirmed all of Bergsmann's results (1965), but added to them the fact that these results could be demonstrated not only in the leukocyte count but in all the other criteria of nonspecific regulation.

Differences between the two sides were most clearly observed in iodometry in the ICV and also in oxymetry of venous blood. There were also side-to-side differences in the electrolytes, blood fats (total cholesterol), and finally also in the immunoglobulins A, M, and G.

Although 30 years previously, a higher antibody titer had already been observed in one-sided inflammatory processes (McMaster and Hudack 1935), but attracted little notice. Only in connection with matrix system research did this earlier observation gain relevance, since it became possible to show that this local increase in antibodies was functionally dependent on matrix regulation.

Table 4 and Figures 12a and 12b show that the initial states and types of reaction vary in the two sides of the body in many chronic illnesses. This can be seen even more clearly in the curves that graph the course of the illness.

During the course of the investigations it became increasingly clear that the matrix system reacts as a whole, but not necessarily in a uniform manner. The less the chronic disease has progressed, the greater these differences in reaction are. Especially in the field of matrix regulation, the time factor, that is, the duration of the stress, plays an extremely important role in spreading the disturbances over the entire organism. This will be discussed later in greater detail.

However, this asymmetry clearly shows that the matrix system has a considerable amount of local autonomy, as was seen earlier in the stimulation threshold experiments, because in the arteries of both sides of the body, the oxyhemoglobin content is equally high, between 96% and 98%. Pischinger and Stacher (Pischinger 1975) were the first to do research comparing arterial oxyhemoglobin (femoral artery) with venous hemoglobin (cubital vein). They demonstrated that the oxyhemoglobin level was the same in the arteries on both sides of the body but was different on each side in the veins. Research by other investigators have confirmed these results both in hemoglobin and electrolyte determinations (Ca, K, Mg).

2 ✦ Testing the Initial State and Autonomic Asymmetry

In the question of whether these processes are under central control or have a more peripheral autonomy, the observations tend to favor peripheral autonomy. This opinion is further supported by the fact that in each case the more seriously disturbed initial states and reaction types are found in the side that is more severely stressed. Unavoidable tissue disintegration, which is found to be the cause of such side-to-side differences, is always involved. At the same time, it is

Table 4

Prot. No.	Time in Hours	% Oxy-Hb right/left	Iodine Consumption Value in mg% right/left
65	0	55/48	103.9/ 103.4
	1	36/33	103.2/ 103.2
	3	38/46	100.2/ 98.2
66	0	50/60	90.7/ 89.2
	1	74/84	90.7/ 86.6
	3	87/90	90.7/ 89.6
71	0	66/83	100.7/ 97.5
	1	66/80	93.3/ 95.7
	3	51/60	94.0/ 91.2
77	0	64/44	82.4/ 77.9
	1	50/55	79.1/ 79.8
	3	64/56	84.3/ 79.1
83	0	42/22	90.6/ 01.0
	1	38/27	87.0/ 85.8
	3	36/27	87.4/ 86.2

Reflex Oxymetry Prot. No. 680314 plo-27

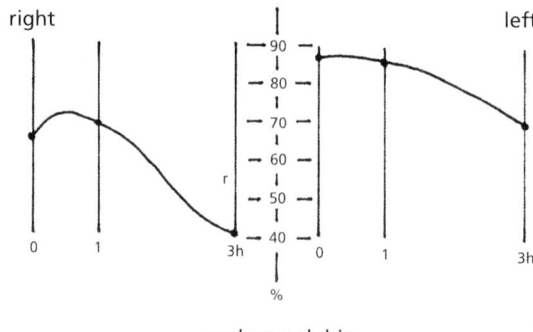

oxyhemoglobin

Figure 12a: Differences in the initial value and type of reaction in oxymetry of venous blood (from the slide collection of Prof. G. Kellner, Histol.-Embryol. Institute, Vienna University).

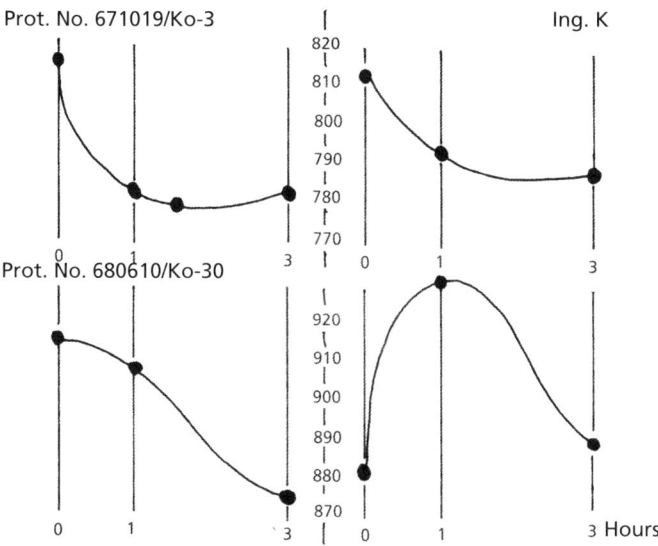

Figure 12b: Differences in initial value and type of reaction in iodometry (from the slide collection of Prof. G. Kellner, Histol.-Embryol. Institute, Vienna University).

irrelevant whether there are inflammatory disturbances (so-called foci), or inorganic stress (e.g., surgical scars with talcum crystal inclusions, war wounds with shrapnel and the remnants of foreign matter, or accident scars containing sand, asphalt, or glass splinters). In addition, we must bring up two further experiments that highlight that peripheral control is more important than central control. Both investigations used cell cultures which allowed the observation of functions without central nervous or hormonal influences.

Kellner (1963) showed that in the matrix system, the acid-base balance is regulated; if the culture medium is acid, a neutral pH is reestablished by the breakdown of fibroblasts, and if alkaline, by fibroblast proliferation. In the same year, McLaughlin (1963) published his findings on embryonic epidermal cell cultures. In vitro, embryonic epidermal cells proliferate in a completely haphazard and undifferentiated way; they only differentiate when mesenchymal cells are added forming a basement membrane, and then the epidermal cells grow in orderly layers. With two completely different goals and methods, these two experiments show that inherent in the matrix system there is a fundamental order, and that it is independent of central influences. For all of these reasons—the side-to-side differences, the level of the threshold stimulus, the localization of the stress, and the cell cultures—it can be assumed that matrix system autonomy actually exists.

However, the existence of functional autonomy does not mean that the matrix system cannot be influenced by other systems. The opposite is the case: the interlinking of the regulatory systems is so interconnected that in healthy subjects, it is impossible to tell which regulatory system actually controls the individual nonspecific functions. This is why even [the field of] physiology passed right by the insignificant extracellular matrix. The lymph nodes and hormonal glands and the CNS are far more imposing organs. Beside them, the extracellular substance is diffuse, hard to understand, and can hardly be recognized as an organ system. It is only possible to tell what the individual systems do by their dysfunction when the immune system is disturbed.

However, even then, a clear classification is not always possible. In cases with these regulatory disturbances, this is shown by measurements of electrical phenomena and skin temperature (R and C measurements according to Kracmar [1961]), decoder impulse dermography or thermoregulatory measurements according to Schwamm (1955). Who, then, is qualified to decide conclusively whether peripheral autonomy or central control predominates? Naturally, based on results of humoral criteria, there is a tendency to give a relatively high ranking to the role of the periphery in electrical and thermal regulatory disturbances. There are also observations as follows: both determinations in the early stages only show disturbances the area of the affected segments. Further along during the course of the illness, these disturbances attack the homolateral half of the body. Disturbances of the electrical and thermal reactions of the entire organism only appear in the late stages. However, even then, the intensity of these disturbances is most evident at the affected segment. It is impossible to imagine that these changes can extend over the site of the stress without the autonomic nerve centers being involved. The extraordinary interlinking of the regulatory system cannot simply be denied just because it creates particular difficulties in reasoning.

The question in this is *how*, that is, along which routes (i.e., regulatory systems) can the disturbances spread across the area of a local disintegration. Remember that every "focus" and every "disturbance field" (*Störfeld*) reaches into the loose connective tissue as the conveyor of nonspecific regulatory information (Pischinger 1954, 1956). According to Pischinger, a local regulatory disturbance spreads mainly as a reflex along the communication pathways of the matrix system. This can only occur if there is a bilateral structure that has a specific independence in the two halves of the body regardless of the wholistic structure that the matrix system has. This hypothetical structure can only be the nervous system, with its graduated structure starting from its synaptic terrain on through the brain stem and midbrain, all the way to the brain itself. A thought should be added to these reflections of Pischinger. This is purely hypothetical, but nevertheless worth closer investigation. It is well known that disintegrating tissue—whether organic or inorganic in nature—creates an acid environment at the site in question. However, as Pischinger (1954, 1956) and Kellner (1963) have shown, acidosis on the one hand (in the Elpimed® test) leads to the release of fibroblasts from their cellular contacts in free blood cell forms (large reticular cells: monocytes, histiocytes, small reticular cells: lymphocytes). On the other hand, acidosis leads to the breakdown of fibroblasts which brings the environment back to a neutral pH value. Since, however, as Heine (personal communication) reported, no significant fibroblast deficiency is found in chronic inflammatory foci as compared to the wider surroundings, the environment must be replacing these local, continual losses. This is conceivable for a local area, and could also be valid for an entire segment. To test this, one would therefore need to determine the concentration of fibroblasts both in a local and a segmental area, as well as in a distant, undisturbed segment. This would also be important for the explanation and acceptance of the effects of the so-called foci. For only in a very low percentage of the affected patients does the focus—the silent, chronic inflammation—have a direct, causal connection with the secondary disease, and this focus is merely a predisposing factor for inflammatory systemic diseases, as classified by Kerl (1930) and Urbach (1935) (Perger 1978).

During further investigation, stresses from toxic heavy metals as well as deficiency states were also recorded. The toxic stress included the heavy metals lead, cadmium, and mercury, and to some extent, nickel. The degree of stress was still subsymptomatic, i.e., there were no unambiguous symptoms of chronic poisoning, and for this reason it is defined as subsymptomatic toxicosis. However, even subsymptomatic stresses can lead to immune disturbances where they are deposited, although these disturbances have less of an effect on the matrix system and more on the enzymatic processes in their formation of immunoglobulins. At the same time, deficiencies were recorded in the heavy metals which are coenzymes (iron, copper, zinc, selenium, and manganese) and in minerals (calcium, potassium, and magnesium). However, it is beyond the scope of this book to go into detail about these research results.

Here, it is interesting to note that immune stimulation (in the Elpimed® test) caused only minimal, and sometimes, no side-to-side differences in focal and disturbance field processes. Toxic states and deficiencies are diffuse in their effect, and cover over the particular local tissue changes with silent, chronic inflammation and scar disturbance fields.

This difference in nonspecific regulatory behavior is remarkable and has an important therapeutic consequence. Surgical removal of foci and disturbance fields without treating the toxic stresses and deficiency states can result in complete failure of the therapeutic method. Here, zinc deficiency has the greatest significance. It signifies that the toxic heavy metals cannot be filtered out, and so wherever they are deposited, they block a variety of enzymatic reactions. If there are deficiencies, this means that other enzyme activities are also inactivated. The enzyme metallothionine is responsible for chelating and filtering out toxic heavy metals, but needs zinc as a co-factor for its activity. Zinc is also needed for DNA and RNA polymerase activity. A disturbance of the RNA polymerases also limits immunoglobulin synthesis, which can lead in turn to a deficiency of urgently needed IgG (Perger

1986, 1987). A residual focal formation after reconstructive surgical procedures (e.g., dry socket processes, tonsillar scar abscesses, etc.) are a common consequence of zinc deficiency, since immune reactions such as T-cell activation and IgG synthesis are delayed and inadequate. The development of these zinc deficiencies and those of other trace elements has been recorded since 1980, and is increasing (Perger 1987).

However, these subsymptomatic toxic states and deficiencies affect the matrix system much less than they do the immune system and other enzymatically controlled biochemical functions. But in order to make sense of indirect influences and retrograde influences on the matrix system, one must not forget the tight interlinking of the regulatory system and its interactions. According to all previous knowledge, the immune system in its functioning is downstream of the matrix system, and yet disturbances in the immune system are not without their retrograde effect on the upstream nonspecific regulatory system.

There is another fact whose cause has not yet been explained. The response of the trace elements (iron, copper, zinc) is generally extremely stable in the immune stimulation test. Of course, there are minimal fluctuations, particularly in allergic reaction forms. With Elpimed® as test stimulus, the total serum protein content often drops by 0.2–0.4% within the first hour. The minimal fluctuations of the trace elements are clearly related to this dilution effect. However, in greatly advanced chronic-progressive inflammation, a high iron and copper lability, and particularly a high zinc lability which greatly exceed the dilution effect are sometimes found. (For a modern presentation of this problem, see Heines 1996.)

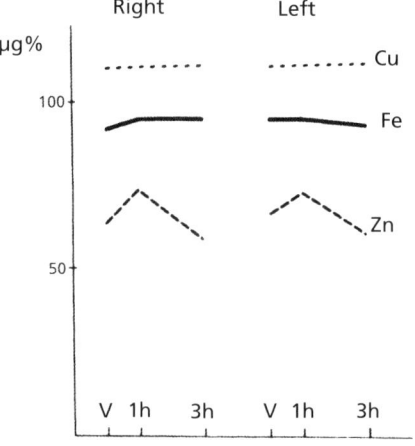

Figure 13a: Stable response of trace elements after test stimulus: seronegative oligoarthritis.

Until now, the reason for these high levels was not clearly defined. It is a fact, however, that they are only found in cases of complete regulatory rigidity of the matrix system, where the immune system's readiness to react is both heightened and disordered. The reason can thus be termed a complete dissociation between the ground system and the immune system. The difference between complete rigidity of the ground system and the over-reaction of the immune system is astonishing. In the most extreme case observed, the IgG rose by 1466 mg% within three hours, which also led to a stronger reaction in an existing primary chronic polyarthritis (this reaction was cushioned with ACTH).

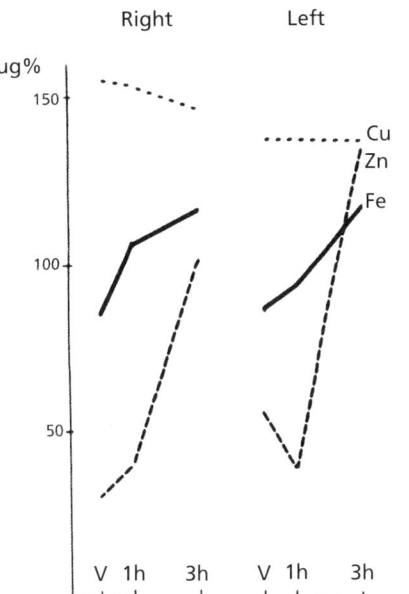

Figure 13b: Labile response of trace elements under test stimulus: advanced seropositive primary chronic polyarthritis.

(Figs. 13a and b from Perger, Öst. Z. f. Stomat. 80 [1990] 289).

When the trace elements show labile behavior, we can assume there is a complete dissociation between the matrix and the immune systems. In none of these cases was it possible to rehabilitate the functions of the two systems. Additionally, all of the cases showed extensive pituitary-adrenal functional exhaustion. Nevertheless, the standard immune suppression methods can lead to a marked amelioration in these patients. This shows that including matrix regulation in the diagnostic procedure makes it possible to determine whether it is still possible to rehabilitate the immune functions and thus can provide a sensible adjunct to both of the therapeutic methods of immune rehabilitation and immune suppression.

Determining the types of reaction and assessing the various stresses points out yet another detail that must be emphasized. The matrix system reacts to all kinds of stimuli in a nonspecific, and therefore, the same way. This is why different types of stimuli can have a cumulative effect. Again and again, we see that by themselves, the individual types of the stimuli are too weak to disturb the function in the nonspecific system, but together they cause a pathological stimulus response. This was obvious in patients where the stresses took the form of several simultaneous chronic inflammations caused by toxicity and abnormalities in intestinal flora.

This cumulative effect is one of the most important findings of research on the matrix system. This effect is also influenced by the time factor, that is, how long the various stresses last.

It is therefore necessary to recognize and take certain facts into account:

1. The ground system reacts as a whole, but not necessarily in a uniform manner.
2. Differences in reactions depend on the localization of the stresses (bacterial and non-bacterial tissue disintegration).
3. The matrix system has a certain amount of peripheral autonomy, but this can be increasingly lost at an early stage of chronic inflammation.
4. Since the matrix system reacts in a completely nonspecific way, a great variety of stimuli and stresses can accumulate and derail its functions; summation effect (chronic inflammation, toxins, deficiencies, intestinal flora abnormalities).

5. The matrix system reacts according to Selye's alarm reaction, and its disturbances according to natural laws are derived from this reaction.
6. There are no means available today that can restore the normal interlinking of the immune and the matrix regulatory systems once they have become dissociated (always combined with exhaustion of pituitary-adrenal function).

3. Extraneural Mechanism for Controlling Immune Processes

Four cellular phases have been identified in immune processes.

First of all, a wall of histiocytes forms around the invasion site of a noxious agent. Second is the microphage phase which follows immediately—this is also still a local phase, but the whole organism reacts along with it somewhat passively. The third phase, the macrophage, is next, and this is accompanied by a full, active participation of the entire organism. Finally the fourth phase, the lymphocyte phase, during which one either recovers from the infection, or it becomes chronic.

With discovery of monocyte factor and the clarification of its position in the immune processes, it became increasingly clear that this factor M is important to activate the macrophage phase. If there is a deficiency of triple-conjugated fatty acids or if they are inactive (i.e., high, but non-fluctuating values in the regulation test), the macrophage phase will be insufficient or not able to be measured at all. This means that this phase will have no noticeable function.

This leads to the logical question of whether there are any similar humoral-activating substances that activate the other cellular phases.

However, this question does not mean that the validity of central nervous regulations should be denied. They are part of an extraneural milieu that, together with central control, is important for immune processes. The precisely formulated question is that—apart from monocyte factor—are there other humoral substances that are also crucial for activating the various cellular phases and, which factors outside neural control are also significant, if humoral substances are out of the question?

It is typical to focus on such thought processes at relatively late point when researching a completely new field such as the matrix system. One enters a labyrinth so to speak, and must feel one's way forward step-by-step, based on the documented facts.

Matrix system research began with the first detail of the discovery of monocyte factor with its completely nonspecific reactions. The clarification of its effects forced a clarification of the histological structure of the intercellular substance. Finally, attention was focused on the different types of reactions after this factor was injected, but the research was complicated by the asymmetrical reactions in the two halves of the body. Parallel to this, recognizing and assessing the causes of immune disturbances in the nonspecific system, as well as the significance of the asymmetries, was a full-time affair. Nevertheless, in 1967 Pischinger had already started to look for further humoral factors that might play a role in the immune function processes. In 1975—at the time that the first edition of this book was published—these efforts had not yet been successful.

It took Pischinger four more years, until 1979, to be able to report the discovery of a second substance that played a role in the cellular process.

The current state of knowledge in reference to the activation of the various cellular phases is given in sequence below.

The first local immune processes are set into motion by a series of tissue hormones (prostaglandins, leukotrines, interferons, etc.). However, the histiocyte wall and microphage phases are not only activated by biochemical but also by biophysical changes—for instance, through the abrupt change in the pH value at the invasion site of a noxious agent—through acidosis and the resulting changes in the cell membranes. This is really quite logical when seen from a biological viewpoint, as Perger (1990) has already described. A humoral process or a biochemical reaction always requires a certain amount of time to create enough of a concentration to have an effect. This wasting of time would be dangerous for the organism. The abrupt change in the biophysical situation at the invasion site leads to a sudden emergency reaction—it immediately sets off the first immune reactions, which are above all, containment measures.

These can be summarized in two phases.

1. releases of large reticular cells from their local bindings in the matrix system; as free cells, these become mononuclear histiocytes and form a wall around the invasion site.
2. changes in the permeability of the capillary walls, followed by the microphage phase.

The microphage phase is not limited to migration of granulocytes to the invasion site. The granulocytes phagocytose the invading pathogens and partly break down under the release of oxidative and proteolytic enzymes, and thus fight the invasion locally. At the same time, the blood serum flows over into the tissue, causing a local edema. This has two effects: the edema further dilutes the noxious agent, and circulating immunoglobulins already available from previous infections can immediately begin to work at the invasion site. For a time, the significance of local edema was controversial, for in it, some researchers saw a danger of pathogens being carried from the local tissue into the entire organism. However, it was finally recognized that this dilution opposed a "high zone" paralysis, that is, an inhibition of the immunity due to a too-great concentration of pathogenic organisms or toxins (Humphrey and Withe 1972). Let us also raise a hypothetical consideration. Changes in pH also allow energy reactions to start. It should not be forgotten that every active process also needs energy. To make this energy, adenosintriphosphoric acid is broken down from the cell membranes, but this requires activation of ATPases. On the one hand, these enzymes depend on an adequate concentration of calcium in the extracellular tissue fluid, and on the ratio of calcium to magnesium. On the other hand—as far as we can tell from the information available today—these enzymes depend on the intensity of the pH fluctuation. A tissue that is already in acidosis (e.g., as in diabetes and/or silent chronic inflammation, poor nutrition, etc.) shows only a relatively minor variation in the pH value, so this minor increase in acidosis has an equally minor influence on the total reaction, including energy release.

All experience indicates that it is this "starter energy" at the onset of every pathological invasion that has major, and even crucial importance for the further course of an illness. In processes involving invasion by a noxious agent, it is almost impossible to make up for a starter energy deficiency later in the course of the disease. The importance of matrix regulation is also underscored by this.

So much for the known and presumed processes for activating the first two local cellular reactions.

However, both central nervous- and organ-associated reactions (CNS, hormone system, lymph nodes), as well as humoral substances needed for activating the macrophage and lymphocyte phases, are responsible for the following phases. Activation of the macrophage phase with all its accompanying reactions is a main theme of this book and was described in detail by Pischinger (1975). In the battle between the healthy organism and the invading noxious agent, the triggering of the humoral-autonomic shock phase, monocytosis, and activation of the entire immune system (transition to the humoral antishock phase = start of the acute phase) only takes place with the active participation of the monocyte factor, and this, in turn, depends on its propagation in the intercellular substance and in the serum.

The increase in the concentration of this factor in the extracellular matrix is pivotal. This is easily demonstrated by the various ways of administering monocyte factor (Elpimed®). The greatest systemic effect follows subcutaneous injection, since a rich supply of soft connective tissue is present there. Intramuscular injection also meets intercellular substance, so there is almost no noticeable decrease in effect. However, there is almost never a detectable systemic effect with intravenous injection, which apparently depends on its dilution in the blood. Injecting it as an intracutaneous irritant, as described by Busch and Busch (1969) certainly has a strong local effect, but there is very little systemic reaction. Giving oral doses makes no sense, because the fatty acid derivatives are broken down—even a normal hydrochloric acid concentration in the stomach hydrogenates the unsaturated bonds, and part of the evidence of the effectiveness of these fatty acids was the fact that after these bonds were hydrogenated, the substance triggered no further biological effects (Pischinger).

Here is a short summary of these effects:

1. triggering of humoral-autonomic shock states resulting in a raised calcium concentration and reduced magnesium concentration in the tissue fluid;
2. increase in the monocyte count and reduction in the lymphocyte count in the blood;
3. increased leukocytolysis rate in the blood;
4. reduction of venous oxyhemoglobin as a sign of increased peripheral oxygen utilization;
5. a shift and increase in the serum β and γ-globulin fraction.

These reactions are obvious in healthy immune function, but gradually disappear in the course of regulatory system disturbances until they are imperceptible, and can even transform into opposite reactions.

The effect is thus particularly clear in delayed recovery after a simple common cold. Often only one, but sometimes 2–3 monocyte factor injections are needed to break through this immune inhibition. A transient fever also appears in a few patients. In most cases the immunity is blocked because the patient attempts to suppress the cold for family or professional reasons. Fever-lowering medications can do this, but only at the price of initiating a humoral shock reaction that persists for 2–4 months.

Similar shock reactions (and these are clearly stronger) occur after skull injuries (contussion, cerebral contusion) and cerebral insults. These shock states can also be resolved by monocyte factor. Obviously, destroyed cerebral tissue cannot be restored, but there are usually edema or ischemic zones around the affected area whose loss of function is still reversible. Correcting this "focal" reaction leads to rapid activation of the function that remains. Naturally, the effect depends on the size of these ischemic or edematous zones, and at first their size cannot be established clinically. This is why there are sometimes astonishingly large remissions and in other cases only minimal improvement. The effect of monocyte factor is greater than intravenous theophylline derivatives. More than 30 years of experience have shown that with monocyte factor there is no risk of causing secondary bleeding. Shock states after fractures, wounds, and burns are also rapidly resolved, so that specific care that is needed can be initiated. The patient recovers more rapidly because the shock is resolved, clearly accelerating the healing process.

All the examples given pertain to completed processes where there is no danger of activating inflammatory processes.

However, chemotherapy also is included in the humoral-autonomic shock theme. Precisely those chemotherapeutic agents that are most effective are the ones that produce such shock states as a side effect. These are often some of the recognized, specific side effects.

The discovery of the alarm reaction in stress by Selye (1953) also has significance for chemotherapy, since every exogenous stimulus—and this includes the use of therapeutic substances—triggers this alarm reaction. However, if the immune functions have been previously damaged, besides the intended specific effect, the undesirable humoral stress reaction is also triggered and often responds with a dysfunctional reaction, depending on the existing immune situation. The allergic reactions are based on the pathological stimulus response to an exudative shock reaction. Toxic consequences are possible if the matrix functions are blocked. This was shown by checking the nonspecific criteria during use of various highly potent chemotherapeutic agents.

Similarly, combining the leading antirheumatic drugs of the phenylbutazone family (which have been on the market for three decades) with the monocyte factor (Elpimed®), reduced the rate of side effects to practically zero. And even today, combining antibiotic therapy with the monocyte factor, which intensifies the specific effects, still has no side effects (allergies, immune suppression, damage to the hemopoietic system). Also, the failure of antibiotics that are effective in vitro but not in vivo can also be remedied by giving monocyte factor at the same time. This avoids the blocking of nonspecific regulation.

Two further observations must be discussed, since they were at first confusing. If a corticosteroid is injected in combination with Elpimed®, the cortisone effect increases significantly. However, if the factor is not injected until 2–3 hours after the corticosteroid, the effect of the corticosteroid on the matrix system is nullified rapidly, i.e., a humoral shock phase (inhibitory phase) is seen again after another hour. A similar situation was observed with alcohol consumption. The hangover following alcohol abuse is relieved by an ampule of Elpimed® injected subcutaneously half-an-hour to an hour after drinking. However, if an attempt is made to avoid the hangover by injecting Elpimed® shortly before imbibing the alcohol, inebriation takes place more rapidly. This undesired effect was observed in two colleagues from California in 1960. A quarter liter of wine after the "prophylactic" injection was enough to intoxicate them both, even though they were both accustomed to drinking wine.

This is evidence that the specific effect of exogenous substances becomes intensified, which is explained by the activation of the primary humoral shock phase.

During acute infections, the use of monocyte factor is therefore not indicated. Its use in chronic systemic diseases is more subtle. It activates the inflammatory processes, but care has to be taken that this activation is not too strong. One injection to test the immune condition is justifiable, but several doses cannot be justified without knowledge of the existing silent, chronic inflammation. Here, activation of the local processes is possible and a reaction in the secondary localization is to be feared. Immune stimulation with the factor for example, permits recognition of scar abscesses after a tonsillectomy, but it is certainly preferable to try to activate a secondary phenomenon with procaine or lidocaine, as described by Huneke (1983).

Monocyte factor is also a valuable in follow-up treatment after eliminating various stresses (foci, toxins, etc.) and after correcting the deficiencies. Lutz (1949) used alternating stimuli to activate the matrix system. First he gave Elpimed® for 2–3 days, followed by an injection with 4–8 IU aged insulin, to activate a nonspecific shock—and repeated both several times in succession. He was able to trigger regulatory blockade with this "alternating therapy" (Pischinger 1975). This method of reactivating nonspecific functions also proved its value in our own setting, but because the small insulin shock caused a weight gain, the method was given up in favor of the readjusting therapies (*Umstimmungstherapien*) described above.

The ideal addition to activate the immune regulatory system was then treatment with lymphocyte factor, which after many years, Pischinger (1979) finally discovered in the venous lymph sinuses.

Factor L was found in lymph node fluid from which the protein and triglycerides had been removed. This residual extract has an absorption maximum of 2,600 Å in the UV absorption spectrum. Adding other substances shifts the absorption curve to 2,700 Å (Pischinger 1979). According to personal information from Kellner, these other substances include, in particular, ATP and zinc.

The UV absorption maximum indicates that nucleotides are involved in these substances—in normal serum the maximum absorption of uric acid is near this figure (2,900 Å). However, there is not yet enough clarity about these substances, so licensing as a drug is not possible at present.

In animal experiments (guinea pigs), factor L changed the blood count by triggering lymphocytosis, which reaches its maximum after about 24 hours. With a high dose of lymphocyte factor in guinea pigs, lymphocytes were washed out of the lymph nodes in such quantities that an almost 100% emptying of the nodes was documented (Pischinger 1979). The emptying takes place in both the cortex and medulla, confirming that both B and T lymphocytes are activated and move into the blood circulation. This result is extremely important for establishing the "starter" function of this factor—it shows that cell-mediated and humoral lymphocyte reactions are set into motion at the same time and to the same degree.

At the same time as the lymphocytosis, the monocytes decrease in the blood, not just a percentage, but also in their total number. This, too, is typical during a normal (spontaneous) lymphocyte phase, and can be taken as evidence that this lymph extract really is the humoral substance that activates this cellular immune phase.

In animal experiments Pischinger (1975) describes the increase in lymphocytes from 28.7%±5.4% to 70.2%±7.8%, and in a second experiment from 49.9%±12% to 68.6%±1% within 24 hours after subcutaneous injection of the extract. At the same time, the monocyte count decreased from 2.4% to 1.1%.

First, Pischinger and the author of this chapter carried out several experiments on themselves to see whether the extract could be used for human beings. Only then was the lymph extract tested on other people. Lymphocytosis appeared in humans, as it had in guinea pigs, but its peak value appeared much sooner, after only 3 hours. After 24 hours, the reaction was fading away in all test subjects.

The extent of the lymphocytic reaction depends on the dose and the individual immune state of the test subject. If the immune state is disturbed, the reaction is often lower or significantly higher than in healthy subjects, but after several injections a more normal reaction returns. The increase in lymphocytes in the differential blood count was noticeable, particularly in older people. We found the greatest increase in a 74-year-old subject after injecting 1 ml of the lymph extract which had a dry content of 0.5%. In this case, the lymphocytes increased from 21% to 64%, i.e., they increased in the blood by a factor of three within 3 hours.

Along with this reaction, there is a simultaneous increase in gamma globulins, which is already evident after 3 hours, but which only reaches its peak value after 2–48 hours. With starch gel thin-film electrophoresis according to Maruna and Gründig (1968), it was shown that it is mainly IgG that increases along with a lesser IgA increase, while both the actual amount and percentage of IgM drops. A few other investigations, which could not be followed up for unrelated reasons, indicated that the T cell population is also stimulated.

Higher concentrations (1.0–1.2% dry weight/ml) often led to a sudden flare-up of a chronic inflammation, with the appearance of fever. Because of this, dilutions first with a dry content of 0.5% and later 0.1–0.2% were produced. Using a dry weight of 0.2%, and by gradually increasing the subcutaneous injections from an initial 0.2–0.3 ml once-to-twice weekly, it was possible to control the lymphocyte reaction in such a way that no general reaction or fever were activated and still achieve the desired clinical effects.

Nevertheless, in severe inflammatory systemic diseases, using solutions with 0.2% dry weight/ml resulted in a rather protracted immunoglobulin reaction compared to reactions from using the initial concentrations of 1.0% or 0.5%.

In the less serious chronic types of inflammation, such as chronic bronchitis, the reaction was of course weaker, but so to speak, regular. However, there were significant differences in the first hours following the injection in patients with multiple sclerosis or rheumatoid arthritis. Moreover, patients with tumors have a rather different reaction.

a) Inflammatory forms of rheumatoid arthritis.

In 24 patients with seronegative and seropositive polyarthritis, injection of 0.3 ml of a 0.2% solution of the lymph extract was followed by a reduction of IgA and IgM of between 8%–17% within three hours, and initially, IgG also decreased by an average of 57%. After 24 hours there was an isolated elevation of IgG which averaged 13% higher than the initial value. The initial reduction in serum immunoglobulins is scarcely a breakdown; rather it is a transfer into the tissues. The subsequent increase in IgG can be correlated with an increased leukolysis rate.

b) Multiple sclerosis

In 22 multiple sclerosis patients, the IgG rose by 10–15% in the first three hours, just above the error limits of the study; the IgM also rose by about –5%, and IgG dropped by about 10%. Only after 24 hours did the average IgG slightly increase by 4%. This minor reaction indicates that the humoral immune reaction is severely disturbed in MS.

c) Malignant tumors

In 55 tumor patients—all following surgical removal of the primary tumor—a decrease in IgM of about 13% was observed within 3 hours; at the same time IgA and IgG rose by about 5%, and after 24 hours they rose by about a further 8%.

These reactions in all three patient groups also are dependent on zinc content. Particularly if there is a serum zinc of < 30 µg/dl, these immunoglobulin shifts cannot be initiated. In addition, the lymph extract is unable to activate the RNA polymerases, since these require zinc as a coenzyme for RNA synthesis. Whether, and to what extent a zinc deficiency may adversely affect leukolysis has not yet been investigated.

However, an increase in the specific antibody titer was also observed. From

1983, due to the theory of slow virus origin of MS, the antibody titers for rubella, measles, Epstein-Barr virus, and subsequently toxoplasmosis, were measured in all MS patients. What is striking is that there were inadequate increases in antibody titers for two or three of these stimuli, with only a few exceptions. The antibody titers are at 1:32 and 1:64 even when these infections had taken place decades before. In healthy people, these antibody titers are only 1:8 or 1:16. With very careful use of the lymph extract (0.1 ml of the 0.1% solution as initial dose with a gradual increase of 0.1 ml once weekly) the AB titers rose slowly to 1:512 and then returned to the normal level of 1:16. The clinical course of the illness stabilized at the same time. This provides additional confirmation of the viral origin of multiple sclerosis—but it also shows that here have usually been several viral stresses, and in some cases, also toxoplasmosis. If this is confirmed by further investigation, it would be an important step in explaining the etiology of this disease.

The effect the lymph extract has of increasing the antibody titer was also confirmed in rheumatoid arthritis patients. In those patients with silent chronic inflammation (foci), the ASLO titer rose to 2,000 IV.

These results clearly show the effect of lymph extract on the B lymphocytes.

However, the effect on the T lymphocytes needs further investigation. In our investigations this was not possible because of unrelated reasons. Nevertheless, some information can be given.

In a 45-year-old patient with breast cancer, the T lymphocyte activity at another site was checked preoperatively. The results indicated total inactivity of this cell population. After 10 injections of increasing doses of a 0.1% solution, a test showed that all sub-groups of T lymphocytes were reactivated, although not completely.

It is not possible, however, to bring about complete tumor regression. This was measured during preoperative preparation in two local patients with local breast cancer recurrences. The tumors did regress, but they did not disappear. In a surgically-treated seminoma with abdominal lymph node invasion and a large metastasis in the left kidney, the lymph node metastases regressed so much that the surgeon could find no indication of lymph node invasion during the kidney metastasis operation. However, the kidney metastasis itself had not been affected by the lymph extract.

A lack of lymph extract forced therapy to be interrupted, and all three patients had a rapid recurrence of the tumors, so rapid that two of the patients died after a short time. Experience with 15 other tumor patients indicates that a certain group of patients need the lymph extract for the rest of their lives—rather like a diabetic needing insulin. There is every indication that patients with an iodine utilization of less than 780 µg/dl are no longer able to synthesize the lymphocyte factor (nucleotide). On the other hand, another group of postoperative tumor patients is able to do this. These are the patients with a high IUV (850–1000 µg/ml), whose reactivity has become rigid. If the reaction capacity of the unsaturated bonds can be restored—and the lymph extract is extremely suitable for this, although other methods also seem possible (e.g., mistletoe preparations)—long-term therapy is not needed. This was first observed in a colleague who had three primary tumors: a bladder papilloma, a seminoma ten years later, and a carcinoma of the larynx two years later); five years following immune activation he had no manifestations of malignancy, but then died in an accident.

The effect of the lymph extract on the T lymphocyte population was further substantiated by the effects in intestinal mycoses. A large percentage of tumor patients suffer from an excessive level of mycosis, having more than 10^6 yeast-like fungi/g stool. Patients without tumors or late stages of chronic-progressive systemic inflammation with those numbers of yeast-like fungi (Candida alb. and other Candida species, Trichosporum spec., Rhodorulata spec., and Geotrichum spec., etc.) can be treated successfully as follows: first, the patient receives the antimycotic agent that is most effective in vitro for 10 days; then 3–4 months of milieu therapy with D-lactic acid and freeze-dried symbiotic bacteria. During this period, the yeasts are reduced to less than 10^2, or they are eliminated completely. However, this therapy does not work with tumor patients and those with late-stage, chronic-progressive inflammatory systemic disease. It is almost certain that fungal immunity depends on the cell-mediated (T-cells) immune performance. When treated with the lymph extract, the intestinal mycoses disappear completely—without additional local regulatory therapy. As regards tumor therapy, the antimycotic effect of the extract also explains its influence on tumor growth—humoral immunity fails because of the well-known enhancement effect. As discussed above, previous experience does not indicate that conservative therapy of malignancies is possible, since massive tumors cannot be made to break down. Surgical intervention to reduce tumor load is still absolutely necessary. However, using follow-up immune stimulation treatment, new angles appear.

Similarly, a surprising effect was observed in schizophrenia. Patients who were about to be institutionalized because of recurring hallucinations were able to return to normal society. The hallucinations disappeared. However, the mechanism of this effect in schizophrenia is utterly uncertain, and given the small patient population, we may not assume it will be effective in all types of schizophrenia. This aspect of the effects of lymph extract effect therefore needs further investigation. The effect on oxyhemoglobin and iodine utilization values is comparable to that of monocyte factor, but—as far as can be ascertained—the effect is not as strong. Here, however, there are no adequate studies using healthy subjects, since the extract was only available in limited amounts. It is also highly regrettable that up to the present, it has not been possible to make even small quantities available for further investigation; the investigations have been stopped since the spring of 1986.

However, these investigations show that using the two substances of the matrix regulation system—monocyte factor and lymphocyte factor—it is possible to treat disturbances of the cellular and humoral immune processes at the right phase, and thus attack disease processes in a more normal physiologic way than was previously possible.

4. Neural Therapy According to Huneke

There are many causes of irritation of the matrix system. As far as can be shown with the testing methods available, there is *almost no disruption* in the organism that does not affect the nonspecific matrix functions. The puncture reaction has already demonstrated this. From a general viewpoint, all noxious stimuli bring the matrix system (mesenchyme, reticulo-endothelial system, soft connective tissue) into a situation of stress. These noxious stimuli include injuries or mechanical disturbances, physico-chemical damage, poisons, and tissue-active hormones. This type of damage enters through the skin (percutaneous, intracutaneous, and subcutaneous), or is intramuscular; via the blood (in intravenous injections); or intradural (flooding the cerebrospinal fluid, Speransky 1950). The disturbance fields (i.e., areas of disturbance, *Störfelder*) appear at these reinforced gates, but only manifest themselves as such when local immunity (protection) has broken down, or in other words, if the disturbance field starts to "spread," i.e., if a distant effect develops. This distant effect can be general or local.

One area of disturbance that usually does not receive enough attention is the digestive tract, whose surface is subjected to an environment of digestion. This surface has an organ-specific epithelium. In addition, the intestinal wall in the mucosa and submucosa is, according to the usual definition, *lymphorecticular* or soft cellular tissue with capillaries and nerve endings, and is rich in extracellular or lymphatic fluid. Nerves and vessels have—please take this comment with a grain of salt—no direct functional contact with the epithelium. The entire extracellular matrix system in its purest form exists in the tunica propria and submucosa of the gastrointestinal tract. Considering what humans expect from their digestive tracts, it is no wonder that the intestinal system is often the *largest area of disturbance,* and—as I have seen in patients myself—that diseases (allergies, eczema, etc.) resist treatment until the disturbances areas with their holistic effects, and last but not least, the intestinal flora and intestinal functions, are cleaned up.

It was only possible to establish disturbances of the intestinal flora with precision after 1982. Only then did it become possible to investigate all four levels of possible disturbances from the fresh, still-warm stool. These were cultures and quantity determination of aerobic and anaerobic microorganisms and fungi, as well as investigations for protozoa and nematodes. These investigations were carried out by Prof. J. Thurner (in Vienna since 1982) and her co-workers, and we are grateful to them.

By the middle of the 1987, about 1,700 complete stool cultures had been examined, and all in all showed a varied picture of intestinal microorganism abnormalities, with consequences for the total immune process. A differentiation must be made between primary and secondary dysbiosis (disturbance of the balance of the intestinal flora). The primary ones are predominant; the secondary ones are the consequence of extra-abdominal immune insufficiency caused by immune disturbances, and are thus more difficult to treat.

Primary intestinal flora abnormalities arise from a variety of causes, mainly as a result of gastric acid deficiency (uncooked food is not disinfected), as a late consequence of severe intestinal infections (e.g., after dysentery, typhus, or food poisoning), after antibacterial therapy (antibiotics, sulfonamides, and imidazole, which damage normal intestinal symbiotic bacteria and alter the intestinal environment due to fungal infestation).

Secondary dysbioses occur in severe disturbances of the total immune performance—in cancer patients, after immunosupressive therapy, in the late stages of inflammatory disease (e.g., in TB, the end stages of primary chronic polyarthritis, and MS), and in inherited and acquired immune deficiency.

The colonies of intestinal flora can have various kinds of disturbances. In 4.6% of those tested, there is a complete lack of Escherichia coli and in 41.7% a complete absence of the lactic-acid-forming Lactobacillus acidophilus and Bacterium bifidum, or there is a major reduction in the counts of these normal intestinal symbiotic bacteria. One of the substances produced by E. coli is Pischinger's monocyte factor, which is important in activating the functions of both Peyer's patches and the intestinal lymph nodes (making the lymphocyte factor).

The lactic-acid-forming Lactobacillus acidophilus and Bacterium bifidum have two important roles with respect to physiology. Previous experience shows that their scarcity leads to inhibition of mineral absorption, particularly of trace elements (especially iron and zinc). These bacteria are also necessary for the maintenance of normal intestinal flora ratios. We have only been able to document the quantitative ratios of these lactic-acid-forming bacteria for about a year, and we found that not only is a complete absence important; even a reduction in these symbiotes is significant.

Besides this deficiency of normal symbiotic bacteria, in both the aerobic and anaerobic flora there is a whole range of pathogenic microorganisms as well as microorganisms that decompose and ferment chyme.

Obligate pathogens (pathogenic E. coli strains, Pseudomonas aeruginosa, α and β hemolytic Streptocci, and hemolytic and non-hemolytic Staphylococcus aureus) were present in 21% of all stool cultures. Aerobic dysbiotic strains were found in 80.1%, and anaerobic dysbiotic strains in 76.8% of the cultures. This means that at the same time mixed dysbioses are uncommonly frequent, usually not only one layer of the microorganisms are affected. It must be emphasized that the patients' subjective complaints are generally more severe with anaerobic dysbioses than those with isolated aerobic dysbioses. This is understandable, since anaerobes make up about 90% of the total intestinal flora.

Yeast-like fungi (mainly Candida albicans, but also Candida spec. and C. parapsilosis, Trichosporum spec., Geotrichum spec., Rhodotorula spec., and Torulopsis spec.) are found in 61.2% of all cultures. The pathogenic effect depends on the quantity present; quantities of approximately 102 fungal microorganisms/g stool are not clinically relevant. Disturbances begin to appear with quantities above 103 microorganisms/g stool. Fungal quantities between <103 and >105 set off disturbances that are considered toxic. In addition, the decomposition products of the partly digested food, the metabolic products of the fungi themselves, and the fermentation of carbohydrates which are easily hydrolyzed, particularly sugar, are all important. If there are above 10^6 microorganisms/g, more and more inflammatory symptoms (colitis) appear. Therefore, it is important for the clinical diagnosis to determine not only the type, but also the quantity of the yeast-like fungi present.

Among the protozoa, infections of Giardia lamblia were surprisingly common. Between 1983 and 1985 a massive wave of these infections spread all across Europe—some of our own patients lived in the most northerly part of Schleswig Holstein—and not less than 32.8% of the cases had Giardia cysts in their stool. Only after the harsh winter of 1986–1987 did the number of positive cases drop down to 13.2%. In contrast, all the other protozoa (the non-pathogenic Entamoeba coli, Isospora belli and Sarcocystis species) are very seldom found (a total of 19 cultures), but from the metabolic point of view, they are as important as giardiasis.

Worms and worm eggs are less often found today. Only 4% of the investigations turned up Oxiuriasis (pinworms in children) and ascariasis (hookworm).

However, all these dysbioses—particularly the predominantly mixed forms—lead to major disturbances in immune capabilities. This has three main important consequences:

1. absorption disturbances (vitamins, minerals, trace elements), in which Giardia is a particularly challenging gastrointestinal parasite,
2. toxins (metabolic products of idysbiosis, decomposition strains, fermentation products),
3. blockage of the abdominal lymph system by toxins and living pathogens.

The intestinal tract can thus be considered the most extensive disturbance field for the regulatory system. As such, it stresses the matrix system through toxins and mineral deficiencies, since it is in the abdomen that this tissue is most abundant and in its purest form.

At present, primary dysbioses can be healed in a few weeks or months by a specific therapy for eliminating the obligate pathogenic microorganisms, the large quantity of fungi, and the protozoonoses, followed by an immediate reintroduction of symbiosis using D-lactic acid, freeze-dried intestinal symbiotic bacteria, and some patience.

However, this treatment is ineffective in secondary dysbiosis, which are immune disturbances whose primary causes are not in the abdomen and which were previously described. These disturbances lead to a spiral of negative effects, because the primary disturbance is intensified by the secondary effects of the dysbiosis. Healing is only brought about with simultaneous, successful immune stimulation.

Correcting the pH, the intestinal flora relationships and intestinal function—as already introduced by Pischinger (1975)—is important for the success of regulatory therapy. At any rate, in chronic diseases the intestinal flora relationships are only normal in 2.5% of the patients; in 97.5% the imbalance is more or less severe. It should not be forgotten that gastric acid deficiency and enzyme insufficiency (for example, following liver and pancreas diseases) must be treated at the same time in order to prevent a general recurrence of the dysbiosis.

Mutatis mutandis—the necessary changes having been made—the same applies to the chronic disturbance fields that are common in daily practice, to ENT, dental, and faciomaxillary areas, to chronic appendicitis, to cholecystitis, and to badly healed scars which can often cause "Huneke's secondary phenomenon" as well as the usual disturbance fields.

Like the entire study of disturbance fields (study of foci), the "secondary phenomena" cannot be understood until one has understood the total matrix system with its three "assisting poles": the nerves (midbrain), the cell (lymphatic tissue), and the hormone pool (adrenals).

The understanding of the phenomena came from (1) the development of serum iodometry and electrometry for examining the processes in the matrix system, (2) the discovery of the puncture phenomenon as evidence of the wholistic reaction of the matrix system, (3) recognition that the matrix system determines the reaction state and the type of reaction, (4) recognition that the matrix system reacts quickly and that the main function of the system is directed towards polarization and depolarization, and (5) that every disturbance field event takes place primarily in cellular connective tissue, which can be considered identical to the matrix system, and that such processes lead to changes of tissue potential which completely involve the nonspecific autonomic system. Thus, if a "scattering" disturbance field is present anywhere, the whole matrix system is disturbed, as our bilateral iodometry tests show, although not to the same extent everywhere. At first, general complaints appear. If an organ is affected by an additional specific noxious event that is so strong that it cannot be overcome or evened out, related local damage appears in, for instance, the liver, pancreas, kidney, lung, gastrointestinal tract, etc., as well as in organ-like connective tissue, such as joints, tendons, and similar organs. H. Eppinger (1949) already discussed these processes: he sees the general disturbance as a disturbance of permeability, with protein passing into the tissues.

If substances that can repolarize—Impletol, or Kofficain, or Xyloneural, that is any of these in addition to Elpimed®—are injected into a "guilty" depolarizing center, the total and general situation can return to normal. This restores the immune capacity as much as possible. Consequently, even overall and local symptoms that are at a distance from the original focus, as well as consecutively altered organ functions, are restored to normal, at least as long as, generally speaking, the normalization of the disturbance field lasts, or general regulation has recovered to the extent that the disturbance field influence—termed false metastasis—can be rebalanced.

According to the measurements made by Professor Kracmar (1961), under stable conditions in a column 9 mm in diameter and 15.4 mm long (in a Pravaz injection):

For Factor M (1.2%), R = 0.95 kilo-ohms and C = 1.9

For µF. Impletol (2%), R = 2.1 kilo-ohms and C = 1.1

For µF. Xylorneural (1%), R = 1.4 kilo-ohms and C = 1.01 ?F.

All three substances have an ergometropic and polarizing effect.

The effect of a *paravenous* injection can also be understood from the same point of view. Marchand classified the perivascular tissue which surrounds vessels as the reticuloendothelial system; it has been described for a long time as *loose connective tissue*, and along with its abundant nerve content, it is a classical autonomic matrix system. A hyperpolarized imbalance here is made into a *depolarized field* by injecting it with Impletol or Xylorneural. There should be no difficulty in understanding the intravenous Impletol injection. Obviously, even if the change in the milieu only lasts a short time, it is sufficient to cause leukocyte reactions through leukolysis, and to alter the energetic state in the pervascular and interstitial tissue through capillary permeability, and thus also to change the energetic state in the matrix system.

From time to time one comes across a version of Huneke's phenomenon where there is a *direct* reciprocal effect between the disturbance field and the distant diseased site, and then one discusses, for example, the possibility of specifically delivering Impletol to diseased areas from the treated disturbance field. I hope I have shown that the secondary phenomenon has its basis in a general shift in the entire biological matrix system, with the bioelectrical (oxydoreductive) potential as the central point, and with all the consequences in blood and tissue. This must also lead to restoration in the diseased point that had the greatest reaction—of course, only as far as is anatomically possible. These are the places that the patient notices a change in the condition.

The secondary phenomenon was objectively elucidated this far by Pischinger in 1961, and noted by Huneke. In 1968, it was examined anew by Stacher (1966), and later it was tackled a few more times (e.g., Bergsmann, see contributions). The upshot was that the explanation of the secondary phenomenon is clearer if the structure of the autonomic field is taken into account—as I demonstrated. From these papers, I would like to extract observations made by Stacher on the bioelectric *effect of scars* in the secondary phenomenon. This deals with bioelectrical measurements in scars. Schoeler, Kracmar, Stacher, and Riedl had already demonstrated that scars have an abnormal potential. Stacher carried out observations on appendectomy scars that were disturbance-free, and others that had a disturbance field effect. He writes the following about these scars: "using a simple tube voltmeter, we measured the skin resistance between a ring-shaped electrode on the lower leg immediately under the knee, and individual points on the scar and the surrounding skin." At a disturbance-free appendectomy scar, "there were no apparent differences between the measured points on the scar and points on the surrounding skin." The values were between 100 and 150 kilo-ohms. "In contrast to this, individual points in the scar which were acting like a disturbance field, had a resistance of up to 1400 kilo-ohms higher than the neutral skin points which were at most one centimeter away."

Stacher also points out: "Afterwards, we took random measurements on a series of scars that gave no clear clinical impression of being disturbance fields, and found such increased resistance more often than we expected. This parallels the histological investigations that Kellner carried out independently on a wide variety of scars, since in a high percentage of the scars met the histological criteria for foci. These scars are therefore to be regarded as potential disturbance fields. Besides this, in our resistance measurements, there was the interesting fact that when the measurements were carried out several times in the same way, they varied depending on barometric influences. The variations were much greater in measuring scars than in simultaneous measurements between the two hands. It can be assumed from this that the disturbed scars, and probably all other disturbance fields, have a 'battery' effect, in the pure physical sense!"

Stacher gives some examples of the coincidence between infiltrating a "guilty" disturbance field with Impletol, etc. and the clinical reaction. However, he does not always find that the breakdown of high resistance is connected with the secondary phenomenon. A late reaction can also take place—1 hour after the injection, in Stacher's example. The term "secondary phenomenon" must therefore be taken with a grain of salt.

Stacher points out that the claim made by W. Huneke was correct, namely, that it often happens that only individual areas of the scar act as disturbance fields, and that if measurements are not possible, one should flood and infiltrate the entire scar. With these and other examples, Stacher confirms that the distant effects of an active disturbance field can be made to disappear when the bioelectrical conditions are normalized. It should not be overlooked that the secondary phenomenon is also the main evidence for the reality of the opposite process, namely the incorrectly-named "seeding" effect, proceeding from altered areas of the matrix system. The previously-discussed research results provide an explanation. This pertains to an energetic holistic reaction that leads to a disturbance field effect that is general or local, and an improvement can occur within seconds of turning off this effect.

To conclude this chapter, it can be pointed out that the substances that can be detected by serum iodometry (Pischinger) allowed the basis of humoral regulation in the autonomic matrix system to be recognized for the first time, as well as quantitatively assessed, and generally explained according to chemistry and mechanism of action.

When the wholistic biological matrix system and its characteristics and effects have been recognized and taken into account, one realizes that almost all theoretical or practical medical problems that can be also considered from this perspective.

To conclude this presentation, here are five brief examples which clarify the role of the wholistic biological matrix system in the organism and medicine using the humoral factor (Elpimed®). Examples concerns (1) cardiac crises, (2) carcinoma, (3) stroke, (4) the manifestations of old age, and (5) the problem of mild X-ray irradiation according to Pape (1948).

The issue of the long series of autonomically controlled functions is raised both in theory and in clinical practice. For our purposes, it is sufficient to emphasize what is important in regulation—that is, the control of the unconscious vital functions, made comprehensible by the term "vital nerves" (Müller 1931). According to biology and general physiology teachings (Bethe 1952), when considering these vital nerves, the focus is the energetic system: the physical and physico-colloid-chemical processes. Connected with these are permeability relationships and bioelectrical potentials. These fields contain the home environment of the tissue respiration with regard to ions, acids, bases, water, and last, but not least, oxygen.

The primary heart problems today are *myocardial infarction* and myocardial damage. As far as their cause is concerned, there are two starkly-contrasting opinions, as I said at the Heidelberg symposium: arteriosclerosis and the myocardial theory. According to the myocardial theory, the cause of the infarct, though complex, is not in the processes that clog the arteries but in the primary damage to the heart muscle of the left ventricle. Kern (1974), together with former authors (Thoma, Eden) and more recent authors (von Ardenne), place the damage of the *muscle fibers* at the beginning of the pathological process. Through further loads (stress and risk factors) that prevent repair or minimize it, the primary damage becomes worse, and finally without any help leads to disaster for the coronary arteries.

The key question is, what prevents the healing and repair of damaged cardiac muscle fibers? The studies by Buchner and co-workers show that *hypoxia* of the myocardium leads to focal-type changes. Kern and the remaining defenders of the cardiac muscle theory also include an inhibition of oxygen utilization as a cause of disturbed recovery. W. R. Hess, H. Eppinger, Sarre, and especially K. Wetzler show that oxygen utilization is a responsibility of *autonomic regulation.* Oxygen utilization is increased by *para*sympathetic and decreased by *sym*pathetic tone. Any disturbance in sympathetic tone must increase potential cardiac muscle damage. It should also be remembered that in the heart, as in other organs, nerves do not directly contact the individual muscle fibers through synapses. Here too, the control takes place through the holistic matrix system. If disturbances in the ubiquitous system affect the entire organism, then the heart cannot be excluded. These disturbances must have the greatest affect where after earlier events they were able to withstand the organism's points of minor resistance (*puncta minoris resistentiae*). The disturbance sources include general factors, e.g., long-lasting physical or psychic *over-stress*. However, the potential causes that are commonly looked for and occur in daily practice should not be disregarded; they can also participate in the complex of infarction causes as local disturbance fields (Perger 1990).

Similar considerations are also valid in the problem of cancer. In cancer patients, serum iodometry always indicates regulatory inhibitions. These are caused and maintained not only by "foci" or disturbanced fields, e.g., in the head area, but also by other disturbed sites in the matrix system, for example, originating in the *intestine,* whose entire mucosa shows the typical structure of a matrix system. Naturally, in individual cases, it cannot be determined whether the tumor caused the disturbances, or the disturbances were already present and created favorable conditions for the pathological cells, or whether the immunity interfered with the formation of "autochthonous" cancer cells (ones that originated where found). In an interesting statement, F. Perger (1990) reported that 25% of all of his patients with "regulatory rigidity" developed a tumor in their later years. In any case, it cannot be ignored that regulatory disturbances are involved in the course of cancer. Merely considering the behavior of the connective tissue in and around the tumor leaves no doubt about this. Here, the condition of normal parenchymal cells includes cancer cells, altered by their own energy system (Ruhenstrodt-Bauer et al. 1966, Seeger 1943).

Many individual questions in this field remain unanswered. There are studies that could be done, for example, to find out whether any and/or which connections exist between leukolysis as an autonomic function and cytolysis according to Freund and Kaminer.

Some important basic points emerge from the above in regards to non-surgical (adjunctive) cancer treatment, besides the very necessary early diagnosis and early surgery. These include: (1) elimination of all disturbance areas (foci) as far as possible, (2) restoration of regulatory and immune capacity through targeted postoperative treatment to activate the matrix system in order to avoid stress and damage from the surgery and massive irradiation, (3) avoidance of general stress, and also stress and damage caused by drugs.

In my opinion, and according to my experience (Pischinger 1966), besides the points mentioned above, a basic rule during concomitant cancer treatment is that the *immune* capacity of the matrix system must be strengthened as much as possible to improve tissue respiration, because this action by itself already can weaken the tumor cells. *Contact irradiation,* that is, either X-rays or radium applied directly to an accessible tumor seems to be the only successful way to directly fight a tumor. If not done this way, more "healthy" tissue will be irradiated, or will be affected by medications such as cytotoxic drugs. One must always remember that every treatment with such measures will cause greater effects and damage to vital immune tissues than would contact irradiation.

Let us add a few more fundamental and important words to the problem of radiation therapy. Many years ago with a large group of patients, F. Perger (1950) was able to show that a radiation effect from a dose that is used for a routine screening chest X-ray, lung X-ray, or an orthodiagram, caused a drop in serum calcium.[5] He recognized that this is a change which has the same significance as the shock effect he found in the course of his regulation studies. Besides this, the

---

[5.] An orthodiagram is a tracing showing the outer contours and the exact size of an organ (as the heart) by illuminating the edge of the organ with parallel X-rays through a small, moveable aperture and making the outer edge of the shadow cast upon a fluoroscopic screen.

normally parallel patterns of calcium and cholesterol values diverge.

Using the weak radiation that Pape himself had used on referred cases, I then studied protein-free serum extract by means of the three-hour spectrographic test. A midbrain irradiation with a skin dose of 5 R causes an increase of ultraviolet-absorbing substances that reach a critical range after just one hour. After 3 hours, they are practically back to their original level. After irradiating the sacral field with 150 R, the spectrograms react differently. In the first hour after radiation, there are no changes in the spectrogram. There is a greater change after 3 hours. This means that irradiation of the midbrain, from the point of view of time, works in the opposite way than does irradiation of the skin. Besides this, there is a difference in the speed with which the reactions appear: following irradiation of the midbrain—reactions occur as soon as one hour, compared to the skin irradiation which, despite the higher dosage, takes 3 hours to react. In the first case, the critical substances are released rapidly; in the second case they disappear; whether they have been oxidized or used up in another way has not been decided. The important point about this weak radiation is that the changes brought about in regulation are so minor that the organism can rebalance them on its own. This treatment is thus like a mild push to the immune system, without producing a blocking effect. Medical practitioners who use this method report time and again that it is one of the best adjunctive treatments in cancer (Zabel 1960, overview Perger 1990). Incidentally, flooding the midbrain-pituitary area with shortwaves, according to Schliephake and Samuels, belongs to the same type of therapy. I attribute most of the effect of this therapy to its effect on the midbrain.

The customary radiation therapy using high dosages does not seem harmless, since it puts considerable stress on the matrix system. Kellner, Picha, and Michalica followed the autonomic matrix status of cancer patients who received long-term radiation therapy with the usual doses (Perger 1990). The follow-ups used serum iodometry (Pischinger 1975). There was a continuous reduction of the IUV in the methanol serum extract. On the other hand, it could be avoided by injecting 1 ml Elpimed® before each radiation treatment. This corresponded to the practical experience that "X-ray headaches" can be prevented with Elpimed®.

These types of treatment show that both the mild interventions in the midbrain. and also those in the periphery, set off a holistic reaction which is expressed in serum iodometry in the hours following irradiation. With the midbrain treatment, the reaction is so minor that it can be rebalanced in a few hours. This *push* to the autonomic matrix system is the chemical expression of the beneficial effects of the therapy. Naturally, favorable effects cannot be obtained at once. All these facts indicate how important it is to not disregard biological medical treatment in tumors. One of these treatments, of which there are several, is Leupold's method, which is not only misunderstood, but actively rejected. I knew about this treatment for a long time, and since reading Leupold's book

(1945), I was able to make sense of it. As I discovered, Leupold's instructions were followed by a group of internal medicine specialists without success. In practice, I learned the technique from the Hainburg Northeast physician, Dr. R. Plohberge (1972), who has mastered the technique and also knows where others have made errors. Naturally, it would be a digression to mathematically prove that not a single salt molecule reaches a cell, with however many cells there are in the body and the use of a saline solution as strongly diluted as that required by Leupold. With Plohberger, I saw what the Leupold therapy can do when it is carried out exactly according to instructions.

It is assumed that this therapy is familiar. Essentially, it consists of a preparatory period of 3 weeks on a carbohydrate-free diet. After this, glucose (*Traubenzucker*) is administered intravenously for one day beginning with large quantities and then decreasing the amount, three times. The final carbohydrate dose is given in the form of bread. Between the carbohydrate doses, the patient is given either insulin or Leupold's salt mixture. The treatment is called "alternating therapy" (*Schaukeltherapie*). The test that Leupold uses to evaluate the therapeutic effect is the systemic cholesterol to sugar and phospholipid test = Ch/Zp.

Alternating therapies are not unfamiliar. Lutz and Pischinger (1949) described one type: Elpimed® is given subcutaneously for 2–3 days. The next day, a subcutaneous injection of 5–8 I.U. aged insulin is given simultaneously with white bread, to exclude an undesired hypoglycemic effect. This depends on the effect of the protein elements in the insulin. Elpimed® is ergotropic, it has an affinity for work; the protein elements orient towards nutrients (tropotrophic). In most cases, regulatory paralysis can be interrupted with this treatment. The same effect is achieved by Leupold's interplay between sugar and insulin.

This method produces surprising results in other areas—not only in the treatment of tumors. Psoriasis is a further indication for using it. Even severe whole-body cases can be healed completely (Leupold 1945, Plohberger 1972). More interesting—but not surprising—are the patients who still have foci or disturbance fields are more resistant during treatment than those who are non-stressed, according to the Leupold and F. Lutz.

There is another whole series of cancer treatments. They are summarized and discussed in a book by Zabel (1960). Two of these treatments need to be emphasized: Gerson's diet and treatment, which focuses on the intestine and liver; and hyperthermic therapy (overview, see Perger 1990). According to the bioelectrical measurements made by Hauswirth and Kracmar (1971), warmth produces ergotropic effects.

An analysis of the proven additional therapies in the treatment of cancer shows, whether or not there is an awareness of it, that success always depends on an activation of the immune performance of the organs.

Removing foci and disturbance fields seems to be essential; but the normal rules must be followed, to avoid exacerbating the tumor situation. The procedure for and during the "removal of the focal lesion" is a surgical procedure, which puts the already-weakened organism in danger. (H. M. Plohberger).

In any case, the intestinal function must be rebalanced by adding symbiotic bacteria, or other appropriate measures must be taken to restore the intestinal environment and intestinal flora (Gerlach 1970, Gerson 1958).

Recently there has been a great deal of talk about the work done by Wolff about using the enzyme product Wobe Mugos® usually together with high, but subtoxic amounts of vitamin A.

Elpimed® is a wonderful adjunct in the treatment of cancer. I personally treated a patient with lung metastases following breast cancer and kept her relatively free of discomfort for 2 more years until the last several days before the end, when morphine had to be administered. Other doctors have reported similar cases also (Sommer 1952). In addition, there are case histories before me from Dr. Karl Singer of Bad Vöslau. After he consulted with Dr. Lutz and myself, all patients, whether with operable or inoperable cancer, were given 1 cc Elpimed® and 500 mcg of Vitamin B 12 every two days. Post- menopausal patients were also eventually given a daily corpus luteum tablet (Oorgamatril). As Singer reports on 40 patients, up to the present 13 have died, including 2 who survived 2 and 7 years, respectively. There are 21 surviving patients, including 8 who survived between 8 and 21 years. Naturally, the major disturbance areas were cleaned up, and a dietary regimen was implemented (for literature see Perger 1990).

Fresh cell therapy and hematogenous oxidation have also been used in the treatment of illnesses. I became acquainted with both methods personally. In experiments with guinea pigs I saw the [cancerous] tissue that had been injected, become necrotic, and broken down. All that remained in the end was a connective tissue scar. It has been reported that the adrenal hormones increase following the injection of cells. However, to me this seems to be a sign of an increased stress load (Selye 1953).

With Wherli, I was able to study hematogenous oxidation more precisely. I found the following: Using Wherli's hematogenous oxidation method, the effects of loading the red blood cells with oxygen seems to be less significant than the simultaneous irradiation with UV light and its influence on the white blood cells (leukocytes). The venous blood is loaded with oxygen in the treatment device, a process that normally takes place in the lungs. Fully oxygenated blood flows in the arteries as long as there is no severe lung or blood damage. It is therefore difficult to see the purpose of loading venous blood with oxygen. Meanwhile, if one follows the changes in the leukocyte survival rate after the blood has been flooded with oxygen and irradiated with ultraviolet light, which seems to be similar to an ozone treatment—as one can tell from the odor—major differences appear in the cell counts between the blood taken from the blood vessels before the treatment and the blood after treatment with oxygen and UV light. The following list has information on

1. 5622 / 3740
2. 4052 / 4155 (K)
3. 7859 / 7055
4. 5317 / 3997
5. 6067 / 4962
6. 5988 / 6155
7. 5483 / 5071 (K)
8. 5903 / 5551
9. 4204 / 3447
10. 1723 / 1601
11. 6038 / 5255
12. 6967 / 6551
13. 9545 / 8763
14. 4388 / 4547
15. 6502 / 4585

this. It concerns blood from patients with a variety of chronic diseases. Two of the samples were of stored blood (K). The numbers before the slash are from untreated blood, and the second numbers following from blood after treatment with oxygen and UV.

After treatment, the blood is reinfused.

The difference between the two mean values (the values before and after the slash) is minus 999.4. The t-difference test shows a probability of error between the two states (mean values) of less than $P = 0.001$, which is highly significant.

The question now is, what causes this major loss of leukocytes? Even Storck (1954) describes leukolysis along with other transitional events in his studies on rheumatism as a regulatory disease. First, it must be remembered that leukocytes remain stuck to the glass walls. However, the significance of this should not be overestimated, as the following observation shows. Experiment by Dr. U: before= 4388; after aerating the blood with oxygen alone= 4547. Thus, the leukocyte count remains virtually the same. Only combining oxygen treatment with UV reduces the leukocyte count to 3557.

So there is only one remaining reason that can explain the drop in the cell count: forced leukolysis. This assumption can be supported by counting the lysis forms before and after oxygen and UV treatment: pre–treatment= 10%, and post-treatment with O2 and UV= –14%.

The spectrogram of the protein-free serum extract changes also. This fact has been established, but there has been no further research regarding its significance.

Havelick's method, which is well-respected in clinical practice, obviously also works according to the same principle. Here, venous blood in an open bowl is irradiated with UV light while swirling it lightly (Sehrt date), and then injected intramuscularly. There is no favorable effect if there is no UV irradiation, or if defibrinated blood or only serum is irradiated and then injected. These facts likewise indicate that there is a reaction between leukocytes and ultraviolet radiation.

In summary, hematogenous oxidation and analogous methods indicate that the favorable effects are a result of regenerating and activating the blood through the white blood cells (leukolysis) and the protein-free fraction of the serum, and not from loading the red blood cells with oxygen. Once again, leukolysis is seen to be an important process for the regeneration of the organism.

5. Therapeutic Consequences of Matrix Regulation Research

The consequences of being able to measure the matrix function processes are far-reaching. However, they are evolutionary, not revolutionary. This is not saying that everything in today's medicine should be turned upside down, as many traditionalists fear.

The course of a disease—as every previous generation of physicians knew—depends not only on the type and intensity of the exogenous noxious influences, but just as much on the existing state of immunity. The only new point here is that the immune state does not depend directly on the immune system; it depends on the matrix system. Nevertheless, it must be pointed out that it has been long recognized that other regulatory systems are involved in immune processes, such as the autonomic nervous and the hormonal system, so that the distribution of tasks in the realm of immunity should come as no surprise. A knowledge of matrix regulations and their measurability does not replace specific diagnostic methods; it expands them, by focusing attention on the body's own irreplaceable immune capabilities. A special opportunity for treating a patient arises when the most recent, response-triggering noxious factors are unknown or have not been substantiated without a doubt. Above all, a knowledge of the matrix functions can clarify the origins of multifactorial diseases more precisely than before. This opens up research into previously unimagined dimensions. It must be mentioned that those who are actively researching the matrix system are also well aware that in spite of considerable progress, a great deal remains to be clarified.

In 1949, Eppinger advocated the view that a disease begins in the extracellular matrix, and only attacks the parenchymal cells afterwards. A specific symptomatology—and thus specific diagnoses—only become possible after there has been cellular change. This occurs at a relatively late stage in a disease, considering that various infections have a relatively long prodromal stage. However, determining the immune status, even in clinically healthy subjects, can give an advance indication of how an incipient infection will be combated and regulated. To examine an entire population this way is a pure utopian fantasy, since everything would collapse because of the expenses and personnel involved. The significant results from previous research, however, make it clear which factors should be eliminated as far as possible to improve the health of the general population. The intimate relationship between matrix and immune function, and the dependency of the specific immune performance on the non-specific regulation process affords us the opportunity of rehabilitating the matrix system and thus normalizing immune function, or at least of approaching this goal. Any delay in the otherwise normal course of immune reactions enables, for example, invasive microorganisms to adapt to the environment of the body and promote chronic inflammation.

The goal of regulatory medicine is to restore normal immune function after treating the specific disease symptoms (acute infections, a flare-up of an inflammatory systemic disease, and after tumor load removal, etc.), to create a positive influence on the further course of the disease. Only in this way can the patient be brought back to a state where recurrences, new outbreaks or metastases can be controlled. As previously mentioned, if the specific noxious factors of a disease are not known or cannot be combatted at the time, rehabilitation of the regulatory system presents the only chance of a normal healing, or at least of improving the course of the disease.

Rehabilitating the immune regulatory system, however, requires a rethinking of the therapeutic process—a rethinking that takes the regulatory system into account.

5.1. Selection of Specific Drugs and Avoidance of Delayed Effects

Specific therapeutic methods or substances are to be selected according to the basic principle of causing the least stress possible to the matrix functions—this corresponds to medical school teaching that methods with the lowest possible rate of side effects should be used. Most side effects are caused because of a severe humoral shock reaction, which primarily encourages allergic reactions. Not only is administering a drug an exogenous stimulus that causes a shock reaction following natural laws; this stimulus also attacks the pre-existing state of immunity. If the state of humoral shock is chronic, the exogenous substance strengthens it further. What initially may be difficult for a physician with a "specific-diagnosis" manner of thinking to understand, and what is important about the matrix system, is that it has only one way of reacting to all stimuli—that is, nonspecifically—and that the different types of stimuli have a summation effect. The matrix system is also incapable of differentiating between friend and foe (Heine 1987), especially where its own cells, the fibroblasts are concerned.

Knowing this, a treatment can be devised in such a way that reduces both the side effects and aftereffects of highly specific medications.

The first example is treatment with phenylbutazone, the leading NSAID (non-steroidal anti-inflammatory drug) for arthritis 20–30 years ago. Apart from its antiinflammatory and antirheumatic effects, it had frequent allergic and hemotological side effects, and so was more or less controversial. Adding the monocyte factor (Elpimed®) to the phenylbutazone injection (Irgapyrin® or Butazolidin®) produced both a marked decrease in subjective pain along with a disappearance of the allergic reactions—we noticed no further complications or hematological disturbances. We select oral analgesics and antirheumatics only if they have a minimal effect the matrix functions, or if possible, no effect at all. Of these, the salicylates belong to the substances that have the least effect on regulation—but they are not always sufficient to achieve the desired specific results. In trials of the matrix system under the influence of medications, ibuprofen and Indomethacin stress the matrix regulation only moderately, so they also are generally acceptable for the treatment of active stages of rheumatic diseases.

In spite of their intensive shock effect on the matrix system, the corticoidsteroids cannot always be avoided. However, corticoidsteroid therapy can be administered with significantly less effect on the matrix system and other regulatory systems (bone metabolism, insulin production, etc.) if the entire daily dose is given between six and eight o'clock in the morning. At this time the natural cortisol level is at its highest, and the drug tolerance is significantly improved. Not only matrix system research, but also classical academic physicians like Bethge (1971) discovered how to use circadian rhythm to avoid unnecessary side effects.

Using acothiaprine and cyclosporin, or aureotherapy for immunosuppression actually paralyzes the matrix system, and activation is hardly ever possible, even to a limited extent. It is only justified when it is no longer possible to rehabilitate the regulatory system. If it is implemented from the outset, when the immune function is first derailed and might be rehabilitated, this type of immunosuppression, in our opinion, is a kind of "resigned" therapy. It is certainly acceptable as the final step—somewhat in the sense of palliative surgery for cancer.

The use of antibiotics affords a further example of how side effects impact the matrix system. It is also an example of how simple methods can often prevent aftereffects.

It is common for antibiotics that are effective in vitro not to be effective in vivo. Once again, the cause was found in the humoral shock phases or blockade of the matrix system. A simultaneous administration of monocyte factor not only prevents the reactions and leads to a full antibiotic effect, and in doing so it also prevents allergic reactions. This solves one of the two problems of antibiotic therapy.

Following antibiotic treatment, patients commonly suffer from a second problem, the disturbance of the symbiotic intestinal microbes with the development of bacterial and particularly mycotic dysbiosis. Since antibiotics—like the cells of the matrix system—cannot distinguish between friend and foe either, not only are the pathogenic microorganisms damaged, but often the normal intestinal symbiotic bacteria are damaged as well. Since antibiotics are obtained from mold extracts or from technical processes with similar properties, they make the intestinal milieu prone to fungal infections. Abnormalities of the bacterial and mycotic intestinal flora also lead to significant disturbances in absorbing vitamins, trace elements, and minerals in most of the patients affected. The resulting deficiency states lead to more consequences of the significant disturbances of the immune cycles: mineral deficiency has a direct effect on the transmitter function of the extracellular matrix, since it directly influences the molecular sieve of the proteoglycans (Heine and Schaeg 1979). Again, a trace element deficiency directly inhibits the enzymes of the neurotransmitters as well as specific antibody and immunoglobulin production.

Dysbioses are an important stress factor for immune functions (Perger 1985), and can be easily prevented after antibiotic treatment by simply giving follow-up D-lactic acid and the important intestinal flora symbiotes (E. coli, Lactobacillus. acidophilus, L. bifidum). This prevents the development of these intestinal problems in the first place. The fixed humoral shock reaction in recurrent systemic inflammation (e.g., inflammatory rheumatoid arthritis, multiple sclerosis, ulcerative colitis, etc.) all too easily leads to an allergic-type reaction in unaffected disease events and to medications. Treating the allergy with calcium and antihistamines is thus indicated in the active stages of these diseases. Doing so often accelerates remissions and prevents side effects from the needed specific treatments. Regarding the liberal use of calcium in exacerbations of MS brings the objection that calcium lengthens the excitation conduction and is therefore undesirable. On the other hand, reducing inflammatory edema and thereby preventing tissue destruction should be seen as more important.

In patients with inflammatory systemic diseases, one repeatedly sees that even minor viral infections can damage the immune capacity to the extent that the regulation against the main disease collapses, and new exacerbations appear. Treatment with human gamma-globulin, a widely-accepted method, can largely prevent this problem. A substitution of this type is very close to knowledge of the intensification of the matrix system disturbance and the delayed immune reaction it causes.

This selection of the most important examples may suffice to make conceivable the possibilities of gently protecting the disturbed immune functions. Even the most highly efficacious specific therapeutic agent secondarily initiates a humoral shock effect that can damage an already-damaged regulatory cycle, and then can intensify this damage still further. Giving due attention to these processes leads to a significantly lower rate of side effects, whether in the form of an allergic or toxic reaction which appears immediately before one's eyes, or in an unfavorable effect later in the course of the disease.

In this connection, it must be pointed out that the preliminary experiments, before a new drug is introduced, are done on healthy animals, and that because of this, such nonspecific side effects are not even noticed as they would be during an intensification of immune disturbance. In spite of their impressive effects on a flare-up of inflammation, the disadvantage of many chemotherapeutic substances is that they can initiate a transition from a step-wise disease process form that can be rehabilitated to one of the chronic-progressive forms that can barely be influenced. In order to keep this transition from occurring, these substances should only be administered when absolutely indicated, whether they are called immunosuppressives or only have that kind of effect.

5.1.1. Rehabilitation of Immune Capacity

Restoration of at least near normal immune functions is—as shown by the 40 years of matrix system research—the most important condition for preventing relapses. An extreme example of this is seen in MS patients who can be treated successfully. Two patients lived for 25 years and 30 years respectively, without exacerbations. After 25 years, one patient developed severe toxoplasmosis followed by an exacerbation; the second patient had a severe psychic shock after 30 years, which set off an exacerbation. This means that the precipitating noxious agents (slow viruses?) had been latent in the body the entire time, but that the immune capacity had been raised to such a level that they were only reactivated under extreme conditions. Issel's astonishing successes in the treatment of hopeless cancer patients, which this author and the chairman of the Vienna Health Council were verified on site, are evidence that rehabilitation of the immune functions must be regarded to have the same value as specific therapy.

However, the right conditions must be created, since it is impossible to achieve such an effect and to rehabilitate these functions without first removing the chronic stresses. These stresses do not usually have a direct causal relationship with the ensuing primary disease but only prepare the environment in which such a disease can establish itself. As Urbach (1935) expressed it, this has to do with predisposing factors, the so-called focal infection in a variety of allergic disorders.

According to the current state of knowledge, these are silent chronic inflammations (previously regarded as "foci" with a direct causal connection to the primary disease), intestinal disturbances (all forms of dysbiosis), subsymptomatic toxicoses and deficiency states. In addition there are abacterial disturbance fields and scars with disordered wound healing, often with foreign body inclusions.

Once again, as with chemotherapy, careful consideration must be given to the disturbed immune function and its low capacity to accept stress. Thus elective surgery such as tonsillectomy and extensive dental restoration are delayed until the capacity of the regulatory cycle to accept stress has been increased by eliminating stresses that can be treated conservatively. This eliminates the complications and surgical incidents that gave the old focal research a bad reputation. After treating the acute stages of inflammatory systemic diseases, the next step is treating those stresses that can be eliminated by conservative methods; only then are the necessary elective surgical procedures carried out (Perger 1987).

5.1.2. Conservative Therapy to Relieve Stress on the Immune Regulatory Cycles

If there are deficiencies in vitamins, trace elements, and minerals, or if toxic stresses are present, surgical procedures to eliminate silent, chronic inflammations often lead to disturbances of the healing process, and thus to the formation of residual "foci." Dry socket, scar abscesses in the tonsillar area, and other problems are so common that the effect of implementing such procedures without preparatory measures seems highly questionable at best. The procedure is not simply done to clear up the foci, but to make rehabilitating the immune capacity possible. The conditions for a normal healing process must thus be created in advance.

These stresses are so intertwined in their consequences, that describing them separately is practically impossible. They result from absorption disturbances through dysbioses and enzyme insufficiencies, through the subsymptomatic toxicoses resulting from disturbed detoxification functions, and through deficiencies of vitamins, trace elements, and minerals.

Deficiency states can be traced back to two causes: to inadequate quantities of food, and absorption disturbances in the digestive tract. Subsymptomatic toxicoses are mainly caused by environmental pollution and a disturbed detoxification function, which also can be caused by a trace element deficiency.

Nutritional deficiencies are mainly due to the poor diets many people have. In the example of zinc deficiency, however, it is obvious that even the best intentions do not necessarily lead to ideal nutrition. Seeling et al. (1977) described the frequency of zinc deficiency in calves that corresponds to human acrodermatitis enteropathica, which is also related to extreme zinc deficiency. Kostolics (personal communication 1979) reported that pigs in mass breeding units had skin diseases that similarly could be traced back to a zinc deficiency. Since zinc is best absorbed from meats, there is no guarantee of an adequate supply from animals raised under such conditions. If we merely limit our consideration to zinc, the consequences are extensive: In zinc deficiency, protein synthesis is inhibited, including DNA and RNA synthesis and immunoglobulin formation, since zinc is absolutely essential for the activation of the polymerases. The detoxification function is also disturbed because the metalothionine responsible for the chelation and removal of toxic heavy metals cannot become sufficiently active. Also, with a greatly lowered zinc levels there is almost always a reduction of the β-globulins and thus a considerable inhibition of detoxification processes.

In addition, zinc is functionally necessary for at least 80 enzymes, including the enzymes for the neurotransmitters and insulinases.

It can thus be seen clearly that a zinc deficiency has considerable negative significance for healing processes. Even if in these cases the immune system is more affected than the matrix system, a zinc deficiency still has the same significance for the rehabilitation of immune functions as the mineral deficiencies that affect the matrix system directly.

In our experience, the most significant mineral deficiency is a calcium deficiency, followed by magnesium deficiency. These have a direct effect on the matrix functions, but are very important for other regulatory cycles.

The transmitter function of the molecular sieve, described by Heine and Schaeg (1979) depends on the size of its pores, and therefore its permeability is dependent on the mineral content of the extracellular matrix; it alters this mineral content to a great extent by taking up or giving off water, which is made possible by exchanging sodium and potassium for calcium and magnesium. With the many biological tasks of these minerals, here, even minor deficits are also important. In addition, minerals participate in the release of energy for starting the immune functions. Energy is obtained from ATP acids by activation of ATPases. All previous experience shows that these ATPases are inhibited by magnesium and activated by calcium, so that energy release in the humoral shock phase is inhibited with increased magnesium in the matrix tissues and activated with the increase in calcium in the countershock phase. Here, the presentation by Heine was followed, in regards to where the energy came from. From our own observations, the reactive behavior of the minerals locked into the various phases of the alarm reaction allowed us to draw conclusions regarding the various degrees of activity of the ATPases.

However, absorptive disturbances in dysbioses and dysfunctions in the digestive tract are more important to deficiency states than is a reduced availability of nutrients. Simply a deficiency of stomach acid leads to severe dysbioses—and no exception has been found yet. Gastric acid is not only responsible for starting the digestion of protein, but also for disinfecting uncooked food. In states of low acidity or no acidity, there is a greater susceptibility to intestinal infection in states. This was again confirmed in susceptibility to intestinal giardiasis. Of 444 patients with this protozoal disease, not a single patient had hyperacidity. Only 51 patients (11.5%) had normal acidity, and 393 (88.5%) had subacidity or no acidity at all (Perger 1988).

These examples already demonstrate the importance of normal gastric acid relationships. Further research on stool cultures of approximately 1,700 patients showed that intestinal microbial balance cannot be corrected without supplemental HCl—there is always a changing picture of bacterial and mycotic dysbiosis; the existing unfriendly microorganisms have barely been eliminated when others establish themselves in the intestine.

There is, however, a point of view that HCl deficiency does not need to be corrected, since the pancreatic enzymes could completely break down the protein in food without the process having been started by the gastric juices. However, this does not appear to be entirely true, since food spends only a short time in the duodenum, and because in dysbioses, there are decay processes. Such considered opinions seem mainly to have overlooked that stomach acid functions as a disinfectant.

Absorption disturbances affect vitamins, trace elements, and minerals. The unfriendly microorganisms particularly yeast-like fungi and protozoa, demand these substances for their own survival; the host organism only gets the now-insufficient leftovers. This then leads to the well-illustrated secondary disturbances of immune and detoxification processes.

The inability to adequately detoxify and eliminate was investigated in detail (Perger and Maruna 1986, 1987, Perger 1987), and in fact, in connection with the toxic heavy metals lead, cadmium, and mercury, and partly with nickel. The conclusion was that inhibition of metalothionine in zinc deficiency almost never leads to a clinical picture of chronic poisoning, but that an interesting picture of enzyme disturbances can appear in the organism because of a depot formation. Maruna and Stipinovic gave the first irrefutable evidence in 1974 when they showed the cause of bone necroses subsequent to fractures was a lead content that was 8.2

times higher in the necroses compared to normally healing fractures. The formation of depots of lead and cadmium in bone leads to disturbances of its metabolism and can lead to severe consequences following a secondary injury such as a fracture. Paradontosis with loosening of the teeth can also often be traced back to cadmium depots in the jaw (Perger 1990). These problems are additional to the already-known damage to renal function (Cd) and to intestinal colic (Pb).

Since in subsymptomatic stresses, unlike in a profound toxicity where a rapid flushing out is necessary, and since chelation has noticeable side effects (including the flushing out of essential trace elements), one can kill two birds with one stone by a combination of supplemental zinc and vitamin C. With this, one normalizes the deficiency state and other functions as well, and slowly, but gently flushes out the toxic heavy metals by activating metalothionine.

A vitamin deficiency can be verified by a test injection. It is remarkable how energized the patients feel after first injection, which consists of a combination of a water-soluble form of vitamin A, vitamin B complex, and vitamin C (*Pancebrin Lilly*). With persistent dysbioses, it is better to give the vitamin combination by injection in order to avoid feeding the dysbiotic microorganisms an optimal supply as well. Later, the combination can be given orally. The initially astonishing activation effect of a vitamin injection gradually decreases, and disappears completely when the body depots are filled.

Deficiencies of calcium, magnesium, or iron are corrected by oral administration. All the commonly used supplements can be used for this purpose.

Simple forms of disturbance in intestinal symbiosis can be resolved primarily with D-lactic acid and freeze-dried intestinal symbiotic bacteria. However, taking the replacement bacteria 1-1/2 to 2 hours before the next meal is recommended because it helps them to revive and adhere to the intestinal mucosa. Taking them in the usual way, just before a meal, often does not succeed because food bolus then carries the microorganisms away with it, and they are excreted in the stool.

High degrees of mycoses and protozoan infections are initially minimized by specific medication (antimycotics or imidazole derivatives). This is followed immediately by the above-described replacement of probiotic bacteria. It is important to accept that even though antimycotics kill the fungi that are present, they do not alter the susceptibility of the intestine to fungi, and that imidazole eliminates not only the protozoa but the friendly anaerobic as well. Thus it is essential to follow this elimination by treating the intestinal environment.

With these measures, correcting deficiency states, flushing out toxins, and normalizing intestinal flora, an important relief of the stress on the nonspecific and specific regulatory systems is produced, and the patient's functioning is redirected toward normality.

5.2. Elimination of Scar Disturbance Fields

The so-called scar disturbance fields are still highly controversial as a stress factor, as is their elimination with local anesthetics, that is, with neural therapy. Not every scar is a "disturbance field" (*Störfeld*) with distant effects. Kellner et al. 1979) called the scars with a disturbance field effect "scars with disturbed wound healing," and supported this with vivid histological pictures of surgical scars with talcum crystal inclusions. In our own field, we also found other inclusions, such as shrapnel and shreds of clothes in war wounds, in accident scars where there were often grains of sand, glass slivers, pieces of asphalt, and so on. These inclusions can only be broken down slowly, or not at all. However, they cause acidosis of the surrounding tissue and thus cause a change in matrix regulation. Like silent, chronic inflammation, they can cause distant effects, and in this, there is an astonishing relationship to acupuncture meridians, which affects the localization of the distant effects.

Infiltrating such scars with local anesthetics (procaine or lidocaine) results in repolarization of cell membranes, and this remains well after the effect of the local anesthetic. Through this, as Kellner (1976) described, blockage of the matrix system is dissolved. This can break down the disturbance, or at least regulate it out. Repolarizing the cell membranes is purely a biological effect, but it is important for starting all succeeding biological functions, including those of a biochemical nature. According to all knowledge to date, maintaining the tissue potential is one of the most important tasks of the matrix substance. The two types of fibroblasts have opposite charges, which maintain the tissue's [electrical] potential.

The local anesthetic effect of procaine and lidocaine is essential for this effect. However, it should be regarded as an additional positive factor that the anaesthetic allows the reactivation of the immune functions to proceed in a pain-free manner. Disconnecting the distant effect of abacterial disturbance fields can also improve the immune state.

5.2.1. Release of Silent Chronic Inflammation (Foci)

Surgical elimination of the silent bacterial processes is only guaranteed to be low-risk and to cause no inhibition or disturbance of the healing processes when the deficiency states have been corrected, the intestinal flora relationships at least generally normalized, the toxins flushed out, and the scar disturbance fields eliminated.

The fundamental effect that the so-called foci have is not caused by their microorganism content. This effect owes very little to the distribution of bacteria and toxins. This distribution only triggered septic or allergic reactions in about 3.3% of cases. In 96.7% of the cases, the actual mechanism is similar to that of disturbance fields that are abacterial. In addition, one must take into account the constant expenditure of energy and substance to wall off and keep the infection localized, which is needed to stem spreading of sepsis. This was shown on 7,148 patients (Perger 1990). The reason for clearing up existing silent, chronic inflammations is not because they might spread, but because they cause energy-consuming limitations and tissue acidosis with increased depletion of fibroblasts.

Despite previous improvements in the immune state by the measures described above, the way patients with foci react is still so disturbed that individual surgical procedures must be scheduled so that these patients are not overloaded by a series of surgical shocks which follow too closely in succession. In the heyday of focal therapy (from about 1910 to 1940), elimination was carried out as rapidly and radically as possible, under the assumption of a direct causal relationship between the focus and the secondary disease. However, this mistaken viewpoint of the focal theory of that time (focal infection or focal allergy and focal toxicosis), often led to significant complications and long-term damage. The overloading of the already severely damaged matrix system due to summation of surgical shock often led to a final blockade of ground regulation and a change to chronic-progressive forms, and also to fatal incidents (particularly in septic forms), because the damaged immune cycle could no longer respond adequately to these stresses.

Any vital interventions therefore must be timed in such a way that the organism has enough time for recovery between the individual procedures. At the same time, a protective therapy is needed that reins in the allergic-type of overreaction, but does not suppress the healing processes.

Aiginger and Neumayer (1951) relate this recovery time to the potential for serum sickness to develop; it can appear until up to 21 days after a foreign protein injection—an organism usually needs this same amount of time to recover from surgery or the shock of an accident. This parallel between two stresses from completely different causes show that matrix functions are uniform in regards to all stimuli, and that they are important as starters of normal or pathological immune processes. (Incidentally, at the time, the two authors were completely unaware of the existence and functions of the matrix system.) This means that the length of time between individual surgical procedures must be at least 4–5 weeks, to prevent overload reactions.

The task of protective therapy is to prevent an allergic-type reaction, so the antiallergic substances, calcium and antihistamines, are used. These do not inhibit the normal healing processes.

The use of corticosteroids in protective therapy is controversial since suppressing the healing processes promotes the formation of residual foci (tonsillar scar abscesses, dry socket, etc.). Riccabona (1955) already pointed out that using antibiotics as protective therapy also can favor allergic reactions which suppresses the desired simultaneous immune stimulation.

The purpose of eliminating silent chronic inflammation is to unburden the ailing organism. A simultaneous activation of the immune and healing processes is an additional effect that is important and should not be suppressed. Only an excessive allergic type of reaction should be kept in check.

5.3. Follow-up Rehabilitation Treatment

The measures described above do not, however, free the ailing organism from its stresses. Except in a few early cases, it takes a long time for the regulatory system to recover—too long, under today's living conditions, to keep new stresses and foci from building up. Our own observations show that spontaneous rehabilitation takes from two to three years. Until then, every infection and stimulus provokes a response, although to a lesser extent, in the regulatory derailment that already exists, and this inevitably causes new foci and chronic inflammations to develop—so the spiral of complicated disease events begins anew, all the way to a fresh activation of the inflammatory systemic disease. It is therefore necessary to rehabilitate the immune functions in order to bring the reactive capacity of the regulatory system back to as normal a state as possible.

It would take us too far afield to describe all possible therapies here, but they all primarily affect the matrix system most intensely, and they partly affect the other systems (immune, autonomic, and hormone systems). Balneology and acupuncture have stood the test of time for over 2,000 years, although their effects on matrix regulation were not known. As practiced by an expert, homeopathy, whose effects can be described as "informative," has also for over 200 years been found to be valid. In addition, there is neural therapy to restore the normal interface potentials, stimulation therapies (nonspecific stimuli, fever therapy, ozone therapy, etc.), and semispecific desensitization with foreign vaccines or autovaccines. In experienced hands they are all capable of restoring the immune functions to normal. These therapies can raise the threshold needed to stimulate a holistic response. In normal people, this threshold capacity of the peripheral regulation is 500,000 microorganisms per vaccination, or a stimulus of equivalent intensity. This stimulus threshold was tested in 917 subjects who had a wide variety of diseases. The stimulus must reach this level before a healthy subject reacts, and then with a brief, general reaction. All the chronically ill subjects react when the stimulus level is only 1/1,000 to 1/10 of this amount, and then with their already disarrayed, holistic reaction.

The choice of rehabilitation therapy can be left to the judgment and ability of the treating physician. In nonspecific response of the matrix system to stimuli, it is better to choose a method one is familiar with rather than unnecessarily to try a special method that this author happens to prefer.

5.4. Success and Failure in Regulatory Therapy

Particularly in inflammatory systemic diseases, when the most recent precipitating noxious agent is not clearly known and thus a specific therapy is not possible, restoring normal immune functions is currently the only way for healing to occur, that is, healing of defects. It is important that potential for rehabilitation still exists, which is not always the case, given our current treatment possibilities.

In addition, allergies with immediate reactions and which cannot be controlled with the usual desensitization using specific allergens, are particularly amenable to treatment from the aspect of a disturbed immune state. In many of our patients, the appropriate investigations found a series of predisposing factors such as chronic silent inflammation, deficiency states, and even heavy metal loading (particularly common with mercury, without this necessarily being a specific allergen). Rehabilitation therapy kept 220 of 260 allergy patients (i.e., 84.6%) free of attacks (Perger 1978).

In episodic forms of inflammatory systemic disease (rheumatoid arthritis, MS, ulcerative colitis, etc.), no further episodes were observed in 82% of the patients observed over a period of 8–30 years. Of 1,136 patients with seronegative oligoarthritis and polyarthritis, only 204 patients had relapses, mainly because new foci or severe acute disease developed; but only 19 of these relapsing patients were subsequently not able to be influenced by regulation. Up to now, 121 patients with multiple sclerosis have been treated, and 99 remained free of exacerbations during the entire observation period. Chronic-progressive diseases present completely different and much less favorable pictures because of the increased risk due to surgical procedures. Many of these patients cannot be cleared up with these therapies because surgical procedures would put them at too high a risk. Among those that could potentially be treated, immunity was restored in only 18%. It did not succeed in 82%. In addition, this is a matter of healing defects, since this treatment cannot reverse irreversible tissue changes.

However, it can be seen from the almost 40 years of matrix regulation research, that it is possible to achieve a genuine healing of chronic diseases which can lead to restoration of the normal harmonious functions of the immune circuits. All experience shows that this cannot be achieved without normalizing the matrix regulation as a "starter" of the immune processes. It is unusual for both the cellular and humoral immune reactions to be simultaneously so strongly disturbed that they cannot proceed normally after the matrix functions have been normalized. Even the pituitary-adrenal axis recovers, as do the vascular nervous functions, when the cybernetic aspect of the matrix functioning is attended to and rehabilitated. At present, it is only impossible to restore these regulatory functions in cases where the matrix system is irreparably paralyzed—above all, in chronic, progressive processes.

In addition, at present, the paralysis of the matrix system can be overcome in some patients with malignant tumors, but the failure rate is higher than in primary chronic (rheumatoid) polyarthritis, and the number of cases is still too small to raise any hopes. However, it is evident that malignancies involve a failure of the entire immunity—paralysis of the matrix functions, and particularly, failure of the cellular immune reactions.

Stating that the healing of chronic disease is only possible by normalizing the immune functions presents no new insights; the old rule that a chronic inflammation only heals through an acute stage is still valid. However, it was Pischinger who laid down the basic principles of how to set this process in motion without incurring great risk to the patient, and how to do this in cases of serious diseases such as systemic inflammations and malignancies. It is up to his successors and students to resolve the many still-unsolved problems, and to push the limits of the ability to treat immune disturbances as far as possible.

The interlinked regulatory systems of the immune system (biocybernetics) will only function normally when the start of these functions occurs in the intercellular substance—with orderly and energetically sufficient intensity and speed.

In conclusion, to avoid misunderstandings, it must once more be emphasized that humoral diagnosis and therapy neither can nor should replace specific diagnosis and treatment—but it presents the opportunity of expanding the horizons of therapy beyond the specific therapy, and should be valued and implemented in equal measure.

There is a particularly important conclusion to be drawn from these investigations. The first reaction at the site of the noxious agent invasion is not only marked by biochemical (humoral) reactions but is also clearly triggered by a pH shift towards acidosis. This biophysical reaction to all exogenous stimuli, not only living agents, is the same. Bacteria, protozoa, and viruses, all set off the same effect as do toxins, trauma, and ionizing radiation. The eminently nonspecific reaction of the precipitation of humoral shock, through the summation of all the stimuli, can thus lead to depletion of matrix regulation, and commonly lead to a pathological response to other exogenous stimuli. To the best of our knowledge, this fact offers us the first concrete opportunity to demonstrate multicausal pathological processes with precision—and this is the most important aspect of the matrix regulation research of Pischinger and Heine.

6. References

Achard, Ch. and E. Feuille. "Sur la resistance leukozytaire," *C. R. Soz. Biol.* 59 (1907): 795.

Adler, E. *Erkrankungen durch Störfelder im Trigeminusbereich*, revised 3rd ed. Medizin Dr. Ewald Fischer: Heildelberg, 1983.

Aiginger, J., and E. Neimayer. "Die Beziehungen der Multiplen Sklerose zu anderen medizinischen Disziplinen." *Klin. Med.* (1951): 2.

Aschoff, L. *Pathologische Anatomie,* 3rd ed. G. Fischer Verlag: Jena, 1913.

_____ and Kiyono. "Ein Beitrang zur Lehre von den Makrophagen." *Verb. d. dtsch. Path. Ges.* 16 (1913): 107.

Auböck, L., "Innervationstypen in der menschlichen Appendix" *Acta histochemica* suppl. X (1967): 225-231.

Baroldi, G. "Histopathological study of the intramural artery vessels in relation to the pathology of extramural corony arteries myocardial damage," *Cardiologia* 41 (1962): 364-380.

Bennel, N. v. c. "Physiology of electronic junction," *Ann. N.Y. Acad. Sci.* 37 (1966): 509-539.

Bergsmann, O. "Asymmetrische Leukozytenbefunde bei Lungentuberkulose." *Wien. Klin.* Wschr. 77/37 (1965a): 618.

_____. "Herdwirkung in der Pulmologie." *Die Therapiewoche* 15/24 (1965b): 1284.

_____ and Dambock, E. "Venöse Oxy-Hämoglobin- und Leukozytenseitendifferenz unter Reizkörperbehandlung." *Beitr. Klin. Tub.* 139 (1969): 295-304.

_____, G. Kellner, G., and O. Maresch. "Synopse zur Frage der biologischen Regulationen, Ärztl." *Praxis XXIII* 933 (1971): 1061, 1193, 1376.

_____. "Tuberkulöse Lungenprozesse und Makroregulation." *Pneumonology* 143 (1970): 247.

_____. "Durchströmungsasymmetrien der Subclaviaarterien in Abhängigkeit von Lungenprozessen," *Wien Z. f. Inn. Med. U. ihre Grenzgeb.* 52 (1972): 152.

_____. "Abühlungstests in der Thermodiagnostik." *Phys. Ther. u. Rehab.* 21 (1980): 661.

v. Bertalanffy. 1952, quoted in H. Heine, "Die Grundregulation aus neuer Sicht." *Ärztezeitschr. f. Naturheilverf.* 28 (1987): 909.

Bessis, M. *Cellules du sang*. Masson et Cie, 1972.

Berthe, A. *Allgemeine Physiologie*. Springer Verlag: Berlin/Heidelberg, 1952.

Bethge, H. "Alternierende Corticoid-Therapie" *Dt. Med. Wschr.* 96 (1971): 1254.

Biermann, H. K, Kelly, Petrakis, Cordes, Forster, Lose. "The effect of intraven. histamin administration on the level of the white bood count in the peripheral blood." *Blood* 6 (1951): 926.

Bircher, F. E. *Autoallergie*. Ott-Verlag, 1968.

Birkmeier, W., and W. Winkler. *Klinik und Therapie der vegetativen Funktionsstörung*. Springer Verlag: Wien, 1951.

Boeke, J. "The sympathetic endformation, its synaptology, the interstial cells, the periterminal network, and its bearing on the neurone theory." Discussion and critique. *Acta anat.* (Basel) 8 (1949): 18.

Bogomolez, H. "Konstitution und Mesenchymen." *Ref. Zentralbl. f. allg. Pathol. Anat.* 35 (1924-25): 375.

Botar, J. "Physiologisch- morphologische Untersuchungen über die Innervation des Nebennierenmarkes beim Hund." *Acta. anat.* 35, suppl. 33 (1958): 1-88.

———. "The innervation of the heart musculature and its changes." *Z. mikr. anat. Forsch.* 70 (1963): 168-214.

Brettschneider, H. "Die Gefäßinnervation als ein Beispiel für die feinere Morphologie der vegetative Endformation." *Anat. Anz.* 113 (1964): 150-171.

Brückle, G. "Beobachtungen an der terminalen Unterlippenschliemhaut des Menschen unter besonderer Berücksichtigung der funktionellen Reaktionen." Lecture at the International Congress for Foci (Herde) Research, Baden by Vienna, (1969a).

———. "Kapillarmikroskopische Untersuchungen über die Wirkung eines Glyko-Peptid- Komplexes auf die geschädigte Gefäßwand." *Die Medizin. Welt* 41 (1969b): 1-19.

Bucher, O. Cytologie, *Histologie und mikroskopische Anatomie d. Menschen.* Verl. Hans Huber: Bern, Stuttgart, Wien, 1970.

Busch, H. J. and Busch, L. Abschlußbericht über Forschungsauftrag d. BM f. Verteidigung, BRD [Final report of the Defense Department research contract], 11/30/69.

———, ———. Mitteilung bei der 30. Annual conference of the Dt. Arb. Gem. für Herd- u. Regul. Frschg. Bad Nauheim., 1979.

Buttersack, F. Latente Erkrankungen des Grundgewebes, insbesondere der serösen Häute. Stuttgart, 1912.

Cajal, Ramon S. "Sur les ganglions et plexus nerveux de l'intestin." *C. R. Soc. Biol.* 39 (1893-94): 217.

Carere-Comes, O. "Über die menschlichen bluthaltigen Lymphknoten." *Folia hämat.* 59 (1938).

Carrel, A., and A. H. Ebeling, *J. of Exp. Med.* 44 n. 2, (1926): 261; *J. of Exp. Med.* 44 n. 3 (1926): 285.

Chwalla and Keible. Quoted in Eppinger, *Die Permeabilitätspathologie.*

Croon, R. "Elektroneural-Diagnostik und Therapie nach Croon." *Physik. Med. Und Rehabil. Jg.* 17/44 (1976): 81.

Diehl, F. "Stundien zur Permeabilität der menschilchen Haut unter verschiedenen Bedingungen," quoted in *Exp. Med.* 100 (1937): 145-191.

Dosch, J. P. *Lehrbuch der Neuraltherapie nach Huneke (Prochain-Therapie),* 12th ed. Karl F. Haung Verlag: Heidelberg, 1986. Contains bibliography.

———. ed. Neuraltherapie nach Huneke: Freudenstädter Vorträge 1971/72. Karl F. Haug Verlag: Heidelberg 1974.

Eberius, E. *Wasserbestimmung mit Karl Fischer-Lösung.* Verl. Chemie: Weinheim, 1958.

Ehrlich, P. *Das Sauerstoffbedürfnis des Organismus. Eine farbenanalytische Studie:* Berlin, 1885.

Eppinger, H. *Die Permeabilitätspathologie als die Lehre vom Krankheitsbeginn.* Springer Verlag, Wien, 1949.

Falck, B. "Observations of the cellular localisation of monoamines by a flourescence method." *Acta physiol. Scand.* 56, suppl. 197 (1962): 1-25.

Feyrter, F. *Über die Pathologie der vegetativen*

nervösen Peripherie und ihrer ganglionären Regulationsstätten. Verl. W. Maudrich: Wien, 1951.

Filatow, W. P. "Die biologischen Grundlagen der Gewebstherapie." *Sowjetwissenschaft, naturw.* Part 1 (1952): 37-76.

Fleckenstein, A. *Die periphere Schmerzauslösung und Schermerzaussschaltung*. Steinkopff: Frankfurt, 1950.

_____. "Bioelektrische und biochemische Primärreaktionen bei der Entzündungsgenese," in *Herderkrankungen Grundlagenforschung und Praxis*. Carl Hanser: München, 1956.

Foulk and Bauden. *J. Am. Chem. Soc.* 48 (1926): 2045. Freund, F., and G. Kaminer. *Biochemische Grundlagen der Disposition für Carcinom*. Springer Verlag: Wien, 1925.

Fundalla, S. G. "Zur Biologie des Mesenchyms." *Synopt. Versuch etc. Hippokr.* 26, 23/24 (1955): 693-697, 740-744.

_____. "Das Herdgeschehen im Wandel der Zeiten." *Physik. Med. und Rehab.* 16/19 (1975): 190-194.

Gabler, E. und Bejdl, W. "Die Allergie der Albinoratten unter Kalbserum-Antigen-Einwirkung." *Z. Immun- und Allergieforsch.* 127 (1964): 184-194.

Gerlach, F. "Biologie der Mykoplasmen und ihre Beziehung zu malignen Tumoren." *Wien. Tierärztl. Monatsschr.* 57 (1970): 232-245.

Gerson, M. Eine Krebstherapie. *Bericht über 50 geheilte Fälle*. Hyperion Verlag: Freiburg/Br., 1958.

Gibian, H. "Beitrag des Chemikers zur Struktur unf Funktionsaufklärung der mesenchymalen Grundsubstanz usw," in *Kapillaren und Intersitium*, edited by H. Bartelheimer and G. Küchmeister. Thieme-Verlag: Stuttgart, 1955.

Gildemeister, F. "Über elektrischen Widerstand, Kapazität und Polarisation der Haut." II. Mitt. *Pflüngers Arch. f. ges. Phys.* 219 (1928): 98-110.

Häbler, C. "Physikochemische Medizin," in H. Schade, *Steinkopf*, Dresden, 1939.

Hauss, W. H. "Über die Enstehung und Behandlung rheumatischer Erkrankungen." *Hippokrates* 32/17 (1961): 678.

_____ and G. Junge-Hülsing. "Über die universelle und spezifische Mesenchymreaktion." *Dtsch. med. Wschr.* 86/16 (1961): 763-768.

Hauswirth, O. *Vegetative Konstitutionstherapie*. Springer Verlag: Wien, 1953.

_____. *Bioklimatologie und Klimatherapie Physiotherapie* 64, n. 3 (1973).

_____. "Die D'Arsonvalisation oder Zieleistherapie." *Der Deutsche Badebetrieb,* 62 (1971): 1-8.

Havlicek, H., quoted in E. Serht, "Elektive ultraviolette Strahlung …" (*see* Serht).

Heine, H. "Der Extrazellumlärraum---eine vernachlässigte Dimension der Tumorforschung." *Krebsgeschehen* 17 (1985): 124.

_____. "Weitreichende Wechselwirkung als Grundage der Homöostase-funktoinelle Aspekte der Neuraltherapie." *Ärztezeitschr. f. Naturheilverf.* 28 (1987): 915.

_____. Schaeg, G. "Informationssteurung in der vegetativen Peripherie." *Z. f. Hautkrh.* 28 (1987): 909.

Heines, J. "Regulationsmedizin-ein Ausweg aus der chronischen Krankheit." Naturamed 11 (1996): 30-39.

Hertting, G. and J. Suko. *Influence of Neuronal and Extraneuronal Uptake on Disposition, Metabolism, and Potency of*

Catecholamines, Perspectives in Pharmacology, A Tribute to Jul. Axelrod. Oxford University Press Inc.: Oxford, 1972.

Hertwig, J. *Die Entwicklung des mittleren Keimblattes der Wirbeltiere,* Jena 1881/82.

Hess, R. W. *Funtionelle Organization des vegetativen Nervensystems.* B. Schwabe Verla:, Basel, 1948.

_____. *Das Zwischenhirn, Syndrome, Lokalisationen, Funktionen.* B. Schwabe Verlag,

Basel, 1949.

Höber, R. *Lehrbuch der Physiologie des Menschen.* Springer Verlag: Berlin, 1920.

_____. *Physikalische Chemie der Zellen und Gewebe.* Stämpfli: Berlin, 1927.

Hoff, F. *Klinische Phyiologie und Pathologie.* G. Thieme-Verlag: Stutgart, 1962.

Holasek and Winsauer. "Pers. Mittlg," quoted in A. Pischinger, 1957.

Horstmann, E. *Lymphgefäßbewegungen.* Farbtonfilm, Kiel: Inst. wiss. Film: Göttingen, 1958.

Huck, W. "Zieglers, Beitr." *Z. pathol. Anat. u.z. allg. Pathol.* 66 (1920): 330.

_____. *Münchn. med. Wschr.* 19 (1920): 535.

_____. *Münchn. med. Wschr.* 20 (1920): 573.

_____. *Münchn. med. Wschr.* 21 (1920): 606.

_____. *Münchn. med. Wschr.* 37 (1922): 1325.

_____. "Das Mesenchym." *Naturwissensch.* 141, (1923).

Humphrey, J. H., and R. G. Withe. *Kurzes Lehrbuch der Immunologie,* 2nd ed. Verlag G. Thieme: Stuttgart, 1972.

Huneke, F. *Das Sekunden-Phänomen,* revised 5th ed. Karl F. Haug Verlag: Heidelberg, 1983.

Ito, Toshio. "Recent advances in the study on the fine structure of the hepatic sinusoidal wall: a review." *Gunma Rep. Med. Sci.* 6 (1973): 119-163.

Jabonero, V. "Die vegetative Peripherie." *Acta neurovegetat.* (Wien) suppl. IV (1955).

_____. "Studien über die Synapsen des peripheren vegetativen Nervensystems." *Acta Neuro vegetat.* (Wien) 29/1 (1966): 111.

_____, M. S. Genis and L. Santos. "Beobachtungen über die cadmiumzinkjodidaffinen Elemente der Vorsteherdrüse." *Z. mikr. anat. Forsch.* 69 (1963): 167-199.

Junge-Hüsling, G. *Untersuchungen zur Pathophysiologie des Bindegewebes.* Dr. A. Hüthig Verlag: Heidelberg, 1965.

Kaiser, E. "Zellatmungsversuch," in Pischinger, *Österr. Zschr. Stomat.* 60 (1963): 294.

_____ and L . "Stockinger. Morphologie u. Biochemie des Bindegewebes." *Münch. Med. Wschr.* 113/10 (1971): 321-333.

Kaufmann, H. P. and H. Budwig. *Fette und Seifen,* 1952.

Keller, R. *Die Elektrizität in der Zelle.* Jul. Kittls Nachfolger Keller u. Co. Mähr: Ostrau, 1925.

Kellner, G. "Die Lymphwege der Menschlichen Milz." *Z. mikr. anat. Forsch.* 68 (1962): 564 .

_____. "Die Wirkung des Herdes auf die Labilität des humoralen Systems." *Österr. Z. Stomatol.* 60 (1963) 312.

_____. "Funtionelle Morphologie der Haut und der Narbe, Ärztl." *Praxis* 18 (1966): 89-105.

_____. "Physikochemische Phänomene bei der Metallimprägnation." *Acta histochem.* suppl. X (1969): 279-285.

_____. Probleme der Wundsetzung und der beeinflußten Wundheilung. Österr. Zschr. f. Stomatol. 66 (1969): 122.

____. "Herdgeschehen und Herdnachweis." *ZWR* 1 (1971).

____. "Zum Konnex von Wundsetzung, Wundheilungsstörung und chronischer Entzündung." *Öster, Zeitschr. f. Somatol.* 70 (1973): 82-89.

____. Klenkhart, E. "Zur Differenzierung der Serumjodometrie," in A. Pischinger "Elektrometrische Titration," *Österr. Zeitschr. f. Erforschung. u. Bekämpfung der Krebskrankheit* 25 (1970): 81-88.

____. "Zur Histopathologie des Störfeldes am Beispiel der Narbe." *Phys. Med. u. Rehab.* 10 (1969): 4.

____. "Wundheilung und Wundheilungsstörung." *Erfahrungsheilkunde* 20 (1971): 173.

____, Krammer H, Seidl, KK: ,Objektivierbare Parameter zur vegetativen. Ausgangslage (Regulationsprufung), Die Heilkunst, 91, 63 (1978).

____. "Wundheilung –Mikrowunde (Nadelstich) chirurgischer Laser," in Laser Regulationstherapie. Otsch. Zschr. Akup 22, 86-95 (1979).

____. "Homöopathie als Arzneimittelreiztherapie." *Öst. Apthotheker Ztg.* 34 (1980): 272.

____, W. Michalica, E. Picha, and P. Placheta. "Versuche zur Verminderung der Belastung weiblicher Tumorkranker durch Strahlenbehandlung." *Z. f. Erforsch. u. Bekämpfung d. Krebskrankheit* 26 (1971): 180-191.

____, A. Stacher, and W. Undt. "Beiträge zur Medizin-Meteorologie 1. Grundsätzliches zur Korrelation von Wetter und Mensch." *Z. angew. Bäder- u. Klimaheik.* 11 (1964): 1-4.

____, ____, ____. "Aus der Bioklimatologie: Zur quantitativen Auswertung von Wettereinflüssen auf den Menschen." *Med. Mitt.-Med. wiss. Abtlg. Schering AG-Berlin* 25 (1964): 23-29.

Kerjaschki D. and L. Stockinger. "Struktur und Funktion des Perineuriums. Die Endigungsweise des Perineuriums vegetativer Nerv." *Z. Zellforsch*, 110, (1970): 386-400.

Kerl, W. Wien. K1. WS. (1930), 1365 and Derm. Wiss. 95 (1932), 1253. quoted in E. Urbach, Klinik u. Ther. d. allerg. Krankheiten (1935).

Kern, B. *Der Myokardinfarkt.* Karl F. Haug Verlag: Heidelberg, 1974.

Kihara, T. "Das extravaskuläre Saftbahnsystem." *Fol. Anat. Jap.* 28 (1956): 601.*

Klingenberg, H.G. and E. Peters. "Der Übertritt von Eiweiß und Wasser in Gewebe unter Physiologischen Verhältnissen." *Wien. Z. Inn. Med.* 30 (1949): 22.

Koch, E. Über leukozytäre Abbauzellen. *Klin. Wschr.* 29 (1951): 474.

Koch, F. W. *Das Überleben bei Krebs and Viruskrankheiten. Das Schlüsselprinzip ihrer Heilbarkeit.* Karl. F Haug: Heidelberg, 1975.

Köhler, U. "Die perorale Strophanthintherapie der Angina pectoris." *Notabene medici* 6, H. 8, Sonderdruck (August 1976).

Kohout, J. "Untersuchungen zur Immunreaktion vom Spättyp bei Erkrankungen der Lunge." *Praxis der Pneumologie etc.* 25/9 (1971): 540.

Kolb, R. "Nachweis von Katecholaminem in den Nerven der Pulmonalisklappe des Meerschwienchens." *Acta neurovegetat.* 29/4 (1967): 540.

____, A. Pischinger, and L. Stockinger. "Ultrastruktur der Pulmonalisklappe des Meerschwienchens: Beitrag zum Strudium der vegetativ- nervösen Peripherie." *Z. mikr.-anat. Forsch.* 76 (1967): 184.

Kölliker, A. *Mikroskopische Anatomie oder Gewebelehre des Menschen.* Leipzig, (1850).

———. *Handbuch der Gewebelehre des Menschen.* Leipzig, 1889/1902.

Komiyra, E., and Mitarbeiter. "Extraktion der neurohumoralen blutregulierenden Wirkstoffe. 2. Mitt. Extraktion von Monopoetin etc." *Folia haematol. Neue Folge* 5 (1961): 328-348.

Kostolics, A: Personal communication, 1979.

Kracmar, F. "Vegetative Konstitution und elektrischer Unfall." *Elektromed.* 6 (1961): 169-174.

———. "Über die Änderung des Polarisationswiderstandes und der Polarisationskapazität des menschlitchen Körpers durch vegetative Pharmaka." *Elektromed.* 6 (1961): 158-159.

———. "Zur Biophysik des vegetativen Grundsystems. Physik. Medizin und Rehabilitation." *Z. allg. Med.* 12 (1971): 120-122. Contains bibliography.

Kraus, F. "Vegetatives System und Individualität." *Dtsch. Med. Wschr.* 48 (1922): 1627.

Kraus, H. "Zur Morphologie, Systematik und Funktion der Lymphbefäße." *Z. f. Zellforsch.* 46 (1957): 446-456.

———. "Besonderheit der Kreislaufsteuerung im (lympho)-Retikulären Bindegewebe gegenüber der Kreislaufsteuerung im kollagenen Bindegewebe." *Anat. Anz.* 109 (1961): 225-230.

Krehl, S., and F. Marchand. "Handbuch der allgemeinen Pathologie." IV, sect. 1, 1924, Etiology of inflammation.

Krogh, A. *Die Anatomie und Physiologie der Kapillaren.* Berlin, 1929.

Kunz, H., and L. Popper. "Zur Frage des Bakterien-Übertrittes aus der Blutbahn in die Lymphe." *Z. Klin. Med.* 128 (1935): 568-582.

Langley, J. N. *La systeme systeme nerveux autonome.* Vigot: Paris, 1923.

Laves, W. "Über Faktoren der Leukozytolyse." *Schweiz. Med. Wschr.* 84 n. 39 (1925): 1097.

Lawrentiew, B.J. "Über die Erscheinung der Degeneration und Regeneration im sympathischen Nervensystem." *Z. mikr.-anat. Forsch.* 2 (1925): 201.

Leak, L. V. "Lymphatic capillaries in tail fin of Amphibian Larva: An electron microscopic study." *J. Morphol.* 125 n. 4 (1968): 419.

Leder, L.D. *Der Blutmonozyt.* Springer Verlag: Berlin/Heidelberg/New York, 1967.

Lennert, H. *Lymphknoten und Dianostik; Bandteil A: Cytologie und Lyphadenitis. Handbuch d. spez. pathol. Anatomie und Histologie,* Bd. I, Teil III. Springer Verlag: 1961.

Letterer, E. *Allgemeine Pathologie, Grundlagen und Probleme.* G. Thieme: Stuttgart, 1959.

Leupold, E. *Der Zell- und Gewebsstoffwechsel als innere Krankheitsbedingung.* G. Thieme-Verlag: Leizig, 1945 (I).

———. *Die Bedeutung des Blutchemismus besonders in Beziehung zu Tumorbildung und Tumorabbau.* II Teil. G. Thieme Verlag: Stuutgart, 1954.

Lickint, F. *Die Leukozytenreaktion nach der modernen Reiztherapie usw.* Inaugural Dissertation, Leipzig, 1923.

Lipp, W. "Studien zur Herzinnervation." *Acta anat.* 13 (1951): 30.

_____ and M. Rodin. "Die adrenergen Nervenplexus in Herzklappen." *Verh. d. Anat. Ges.* 1967, 121/ Ertg. 83 (1968).

Löwerstein, W.R. "Permeability of Membrane junction." *Ann. N.Y. Acad. Sci.* 137 (1966): 441-472.

Lumière. *Grundlagen einer neuen Humanmedizin.* Verlag für Medizin: Leipzig, 1927.

Lutz, F., and A. Pischinger. "Über einen neuen Faktor im tierischen Blut usw." *Wien. Med. Wschr.* 99 (1949): 437.

Maehder, K. "Über den Nachweis der perlingualen Strophanthin-Resorption mittels Isotopen." *Med. Klinik* (1955): 104-105.

Maillet, H. "Modifications de la technique de Champy au tetraoxyd l'osmiumiodide de Potassium. Results de son application à l'etude des fibres nerveuses." *C. R. Soc. Biol.* (Paris) 153 (1959): 939.

Maresch, O. "Physikalische Beurteilung der Heilwässer." Vortrag Symposium über Bädertherapie. May 1970 in Baden-Wien.

Maruna, R.F.L., and E. Gründig. *Wien. Med. Wscher.* 117, 903 (1967); *Wien. Med. Wschr.* 118, 724 (1968).

_____. Stipinovic, G. "Über den Bleigehalt der Oberschenkelknochen von Unfallpatienten im Raume Wien." *Wien. Med. Wschr.* 124 (1974): 616.

McLaughlin, C.B. *Symp. Soc. Exper. Biology* 17, Cambridge, 1963; quoted in F.M. Lehmann, *Biologie,* n. 24 of the series of the Bez. Ärztekammer Nordwürttemberg, Verl. Gertner, Stuttgart, 1976.

McMaster and Hudack. *J. Exp. Med.* 61 (1935): 783.

Maximow, A. "Bindegewege und blutbildends Gewebe," in Handbuch d. mikroskosp. *Anat. d. Merschen Bd.* II/1, Springer Verlag: Berlin, 1927, 232-583.

Meier, A., and H. Gottlieb. *Experimentelle Pharmakologie.* 8th ed. Urban und Schwarzenberg, 1933.

Metschinikoff, E. "Die Lehre von den Phagozyten und deren experimentelle Grundlagen." *Handb. d. pathog. Mikroorganismen.* W. Kolle und A. v. Wassermann. G. Fischer: Jena, 1913.

Meyling, H.E. *Das periphere Nervennetz und sein Zusammenhang milt den ortho- und parasympathischen Nervenfasern. Acta neuroveg.* Suppl. IV: "Die neurovegetative Peripherie" (38-63), Springer Verlag: Wien, 1955.

Molenaar, H., and D. Roller. "Die Bestimmung des extrazellularen Wassers beim Gesunden und Kranken," quoted in *klin. Med.* 136 (1939): 1.

Möllendorff, W. von. *Handbuch der mikr. Anat. d. Menschen,* Bd. II. Berlin, 1927.

_____. *Lehrbuch der Histologie und mikr. Anatomie d. Menschen.* G. Fischer: Jena, 1943.

Müller, L.R. *Die Lebensnerven.* Berlin, 1931.

Nägeli. Quoted in Leder, 1967 (*see* above).

Neuberger, F. "Zur Objektivierung endo- und exogener Einflüsse auf den Tonus des veget. Nervensysstems." *Acta otolaryngol.* 51, 332-346.

Neuberger, F. "Zur Objektivierung endo- und exogener Einflüsse auf def den Tonus der akustischen und/oder vestibularen Sensorik." *Monatsschr.f. Ohrenheilkunde und Laryngo-Rhinologie* 94 (1990): 262-274.

Pape, R. "Ergebnisse und Fragen der

Röntgen-Schwachbestrahlung." *Strahlentherapie, Sonderband* 35 (37.Jhg. München (1948): 116.

Perger, F. "Untersuchungen über den Wirkungdmrchanismus der hochmolekularen Fettsäuren im Elpimed bei parenteraler Zuführung." *Med. Klin.* 51 n. 31 (1956): 1299.

_____. "Die Elpimedreaktion zur Erfassung der vegetativen Grundsituation mittels der Blutkriterienbestimmung als Ganzheitstest." *Öst. Z. f. Stomat.* 60 n. 11 (1963a): 440.

_____. "Problematik der medikamentösen Therapie bei Herderkrankungen." *Ärztl. Praxis* XV/47 (1963b): 2596.

_____. "Die Bedeutung der Grundregulation. Die therapeutischen Konsequenzen der Grundregulation." *Erfahrungsheilkunde* 21 (1977): 140.

_____. *Erfahrungsheilkunde* 30 (1981): 39.

_____. "Über den derzeitigen Stand der Herdforschung." *Erfahrungsheilkunde* 26 (1977): 140.

_____. "Chronische Entzündung und Carcinogen aus der Sicht des Grundsystems." *Wien. Med. Wschr.* 128 (1978): 31.

_____. "Multiple Sklerose und Herdgeschehen." *Wien. med. Wschr.* 126 (1976): 283.

_____. *Öst. Z. f. Stomat,* 60 (1963): 440.

_____. "Die Bestimmung der vegetativen Reakionslage und Reaktionsweise bei Multiple Skelrose-Kranken." (Allergiebeilage) *Dt. Med. Wschr.* 81 (1956): 342.

_____. "Herdtherapie-Erolg und Mißerfolg." *Erfahrungsheilkunde* 27 (1978): 805.

_____. "Extranervale Steuerungsmechanismen." *DZA* 24 (1981): 81.

_____. "Sinn und Unsinn von Herdsanierungen bei Erkrankungen des rheumatischen Formenkreises." *Rheuma* 4, I (1981).

_____. "Immunstimulation über das Grundsystem mit körpereigenen Stoffen." *Erfahrungsheilkunde* 35 (1986): 146.

_____. "Regulationsstörungen und Darmkeimverhältnisse." *Erfahrungsheilkunde* 34 (1985): 812.

_____. "Klinische Aspekte des Zinkmangels beim Menschen." *D. prakt. Arzt* 40 (1986): 1591.

_____. "Die Vor- und Nachbehandlung von Herdsanierungen." *Neuromed* 2 (1987): 12, 68.

_____. "Das Zusammenspiel zwischen den Regelsystemen der Abwehr und seine Störungen." *Erfahrungsheilkunde* 36 (1987): 566.

_____. "Unterschiedliche Entwicklung der Schwermetallbelastungen (Pb, Cd, Hg): und ihre Therapie." *Ärztezeitschr. f. Naturheilverf.* 28 (1987): 774.

_____. *Fragen der Herderkrankung. D. Zahnärztekal.* 1988, S. 23-38. Verlag C. Hanser: Müchen, 1988.

_____. "Klinik der Lambliasis intestinali und ihre Verbreitung in Mitteleuropa." *Naturamed* 3 n. 1 (1988).

_____ and R. Maruna. "Zur Frage der subsymptomatischen Schwermetallbelastung beim Menschen (Pb Cd, Hg)." *Erfahrungsheilkunde* 35, 316 (1986).

_____, _____. "Ergebnisse von Blut-, Harn- und Haaranalysen auf Schwermetalle." *Dt. Z. f. biolog. Zahnmed.* 3 (1987): 107.

Perger, F. *Kompendium der Regulations pathologie und therapie.* Sonntag: München (1990).

Peterson, H. "Über Methoden zum Studium

des Knochens." *Z. wiss. Mikr.* 43 (1926): 355.

Petueli, R. *Biochemische Untersuchungen zur Regulation der Dickdarmflora des Säuglings.* Verlag Nortring der wissenschaftl. Verbände Österr.: Wien, 1957.

Pfeiffer, H. *Wien, Klin. Wschr.* 30 (1921): 363.

_____. Standenath, Fr. *Z. f. d. ges. exp. Med.* 37, 184.

Pichlmayr, R. und Mitarb. "Gewinnung von heterologen Immunseren gegen menschliche Lymphozyten." *Klin. Wschr.* 46 (1968): 249-258.

Pischinger, A. "Die Lage des isoelektrischen Punktes histologischer Elemnte als Ursache ihrer ver Schiedenen Fäbbarkeit Z. *Zellf. u. mikr. Anat.* 3 (1926).

_____. "Neue Beobachtungen zur Aufklärung der Moorwirkung." *Die österr. Moorforschg.* 4 (1953).

_____. "Schicksal und Wirkund körperfremden Gewebes im Organismen." *Die Medizinische, Welt,* 23 (June 6, 1953).

_____. *Pathologische Grundlagen und Probleme der Herderkrankungen. Nauheimer Tagung der DAH.* Hanser Verlag: München, 1954.

_____. "Untersuchungen am Blute nach künslicher Oxygenation." *Müchn. Med. Wschr.* 96 (1954): 879.

_____. "Das System des Unspezifischen und seine Bedeutung für das Herdgeschehen," in *Herderkrankungen. Grundlagenforschung und Praxis.* Hanser Verlag: München, 1956.

_____. "Neue Auffassungen über das Vegetativum, seine Oragnisation und Bedeutung für das Herdgescgehen." *Österr. Z. Stomatologie* 53 (1956): 621-629.

_____. "Die Bedeutung hochmolekularer ungesättigter Fettsäuren im Blut." *Therapiewoche* 7 (1957): 397.

_____. "Das Schicksal der Leukozyten." *Z. mikr.-anat. Forsch.* 63 (1957): 169-192.

_____. "Uuber die Zellen des weichen Bindegewebes," *Wien. Klin. Wschr.* 71 (1959): 73-77.

_____. "Über die vegetativen, insbesondre humoralen Grundlagen des Herdgeschehens." *Ärztl. Praxis* XIII (1961): 249-251.

_____. "Der Monozyt des Blutes." *Symp. Biol. Hung.* 2 (1961): 27-36.

_____. "Über die Organisation des lymphatischen Gewebes." *Z. Zellforsch.* 60 (1963): 893-908.

_____. "Die Objektivierung des Sekundenphänomens (F. Huneke)." *Physikalisch Diätetische Therapie* 6 (1965a): 1-6.

_____. "Aussprache zu Hamplick H.: Cephakeafontalis-Galeatomie (Vortrag)." *Wien. Klin. Wschr.* 77 (1965b): 912.

_____. "Krebs und Abwehreinrichtungen des Organismus." With G. Draczynki und G. Kellner, *Serumjodometrie. Krebsarzt* 21 (1966): 297-311.

_____. "Zur Grundlegung unspezifischer Behandlungsweisen." *Physikalische Medizin und Rehabilitation* 9 (1968): 7-12.

_____. "Über das vegetative Grundsystem." *Physikalische Med. und Rehabilitation* 10 (1969): 53-57.

_____. "Die Grundregulation." *Erfahrungsheilkunde* 20 (1971): 301, 363; 21 (1972): 33.

_____. *Erfahrungsheilkunde* 28 (1979): 317.

_____. "Objektivierbarkeit der Neraltherapie mittels Serumjodometrie sowie elektrische Mssungen." Frühjahrs-

Symposium 1974 d. Österr. Gesellschaft f. Neuraltherapie (Fieberbrunn): (GEBRO).

_____. "Das Konzept des 'Ganzheitssystems' als Weg zum Verständnis der Neuraltherapie." *Die ärztiche Fortbildung* 1 (1975): 27-30.

_____. *Das System der Grundregulation,* K.F. Haug, Heidelberg, 1975.

_____. "Die humoral-zelluläre Reaktion des lympomonozytären Systems." *Erfahrungsheilkunde* 28, 317 (1979).

_____. Stockinger, L. "Die Nerven der menschlichen Gingiva." *Z. mikr.-anat. Zellforsch.* 89 (1969): 153-181.

Plenk, H. jr. and Raab, H. "Die Nerven der menschlichen Gingiva." *Z. mikr.-anat. Forsch.* 81 (1970): 473-491.

_____. "Die Nerven der menschlichen Gingiva (II. Mitteilung: Weitere histochemische und elektronenmikrokopische Befund)." *Z. mikr.-anat. Forsch.* 81 (1970): 473-491.

Plohberger, H. M. "Karzinom und Herdgeschehen." *Österr. Z. f. Erforschg. und Bekäumpfung der Krebskankheit.* 27. Jg. H. 3, 63-69 (1972).

Plohberger, R. and F. Kracmar. "Krebsbehandlung nach Prof. Leupold im Spiegel der Biophysik." *Erfahrhk.* 19 (1970): 458.

_____. "Krebsbehandlung nach Professor Leupold. *Krebsgeschehen.* (1971): 111-118.

Popp, F. A. *Biophotonen. Schriftenreihe Krebsgeschehen Bd. 6,* 2nd ed. VfM Dr. Ewald Fischer: Heidelberg, 1984.

Raab, H. "Die klinische Bedeutung des Nachweises von adrenergen, autonomen Nerven im menschlichen Zahnfleisch. *Österr. Z. Stomat.* 67, 381-390 (1970).

_____. *Die vegetativen Grundlagen dentogener Herderkrankungen.* W. Maudrich: Wien, 1972.

Ratzenhofer, M. "Morphologie und Bedeutung der Funktionsstörung des Mesenchyms nebst Beobachtungen über Veränderungen am Gefäß-Nervengewebe beim Karzinom." *Wien. Med. Wischr.* 100 (1950): 646-652.

_____. "Innervationsstudien am Magen. 10." *Tag. Der Österr. Gesellsch. f. Chirurgie.* 6 (1969): 26-28.

Rebuck, J.W. and J.H. Crowley. *Ann. N.Y. Acad. Sci.* 59 (1955): 757.

Reichert, C. B. "Vergleichende Beobachtungen über das Bindegewebe und die verwandten Gebilde." Dorpat 1845, S. 168.

Reiser, K.A. "Über die Innervation der Hornhaut des Auges." *Augenheilk.* 109 (1935): 251.

v. Riccabona, A. "Kritik der Handerkrankungen vom HNO-Arzt." In Krit. *Betrachtung des Herdgeschehens* (DAH-Kongr. 1954), Verlag C. Hanser: München, 1955.

Ricker, G. *Pathologie als Naturwissenschaft.* Berlin, 1925.

Rilling, S. *Biotonometrie, Grundlagen und Anwendungen.* Stuttgart, 1971.

v. Rindfleisch, E. *Elemente der Pathologie.* Leipzig 1869.

Rohr, K. "Blut-und Knochenmarksmorphologie der Agranulozyten (Ergebnisse fortlaufender Sternalmarkuntersuchungen)." *Fol. hämat* (Lpz.): 55 (1936): 305-367.

Rokitansky, C. v. *Handbuch der pathologischen Anatomie.* Wien, 1846.

Rothmund, W. "Neue Aspekte zur Infarktforschung. Ein Diskussionbeitrag." *Zeitschrift f. Allgemeinmedizin (ZfA)* 52, n. 16 (1976): 952-960.

Rudich, A. "Probleme der oralen Strophanthinbehandlung." *Physik. Med. und Rehabilitation* 13, n. 7 (1972): 206-211.

_____. *Das Menschliche Knochenmark.* George Thieme-Verlag: Stuttgart, 1960.

Ruhenstroth-Bauer, G., Fuhrmann, Bey, Hertel. "Änderung der Membranladung von Rattenleberzellen usw." *Klin. WSchr.* 44 (1966): 39.

Rumler, K. "Der Säure-Base-Haushalt im Rahmen der Gesetzmäßigkeit der biologischen Regulation." *Ärztl. Praxis* 23 (1971): 420, 481, 541.

_____. "Beeinflussing der Regulation durch Ernährung." *Erfahrungsheilkunde* 21 n. 9 (1972): 269-276.

Ruska, H., and C. Ruska. "Licht-und Elektronenmikroskopie des peripheren neurovegetativen Systems in Hinblick auf die Funktion." *Dtsch. med. Wschr.* 86; 1967 and 1770 (1961): 36, 37.

Salzer, G. M., L. Stockinger, L. and W. Zenker. "Die Ultrastruktur des juxtaoralen Organs der Ratte." *Z. Zellforsch.* 62 (1964): 829-854.

Sato, Itio. "Studien über die Antigenität der monozytogenen Substanzen im Blut." *J. Chosen med. Assoc.* 29: 29/11. German summary 328-329. (1939) Japan.

Schabadasch. "Intramurale Nervengeflechte des Darmrohres." *Z. Zellforsch.* 10 (1930): 320-385.

Schade, H. *Die physikalische Chemie in der inneren Medizin,* 3rd ed., 1912.

Schaffer, J. *Lehrbuch der Histologie und Histogenese.* W. Engelmann: Leipzig, 1922.

Schauenstein, E. *Öster, Chemiker-Zeitung* 52 (1951): 28.

Shauenstein, E., Pibus, B. "Nachweis von Lipo-Peroxyden im menschlichen Serum. Wien." *Klin. Wschr* 68 n. 19 (1956): 376.

_____. "Autoxydation of polyunsaturated esters in water: chemical structure and biological activity of the products." *J. of Lipid Research* 8 (1967): 417-428.

Schilling, V. *Pratische Blutlehre.,* 8th/9th ed. Jena, 1938.

Schliephake, E. *Kurzwellenentherapie,* 5th ed. Piscator-Verlag: Stuttgart, 1952.

Schoeler, H. "Über elektrische Widerstandsmessungen an biologischen Geweben insbesondere an der Haut. Tgs-Bericht d. intern." *Gesellsch. f. Neuraltherapie n. Huneke, Freudenstadt* (1960): 91-101.

Schölffler, H, H. "Zur Ätiologie des Herzinfarkts."*Erfahrungsheilkunde* 1 (1976): 6.

Schröder, H. J. "Gesetzmäßigkeiten bei der Nekrobiose und Autolyse der weißen Blutzellen und ihre biologische Bedeutung." Habil.-Schr. Med. Fak. d. Univ. Hamburg, 1959.

Schröder, H. "Über elektrische Widerstandmessungen an biologischen Geweben insbesondere an der Haut. Tgs-Bericht d. intern." *Gessellsch. f. Neuraltherapie n. Huneke, Freudenstadt* (1960): 91-101.

Schoeler, H. H. "Zur Ätiologie des Herzinfarkts." *Erfahrungsheilkunde* 1 (1976): 6.

Schröder, H. J. "Gesetzmäßigkeiten bei der Nekrobiose und Autolyse der weißen Blutzellen und ihre biologische Bedeutung." Habil.-Schr. Med. Fak. d. Univ. Hamburg, 1959; bibliography included.

Schröder, H., and J. Schröder. "In vitro-Untersuchungen über die Wirkung des Zigarettenrauchens auf die Vitalität der Leukozyten." *Fol. hämat.* 87 (1967): 190-198.

Schröder, J. "Über gesetzmäßige Veränderungen der weißen Blutzellen in hypotoner Flüssigkeit." *Z. Zellforsch.* 46 (1957): 300.

_____. "Der Einfluß blutchemischer Veränderungen auf die Fermentaktivität der Leukozyten," *Acta hämat.* 19 (1958): 156.

_____. "Titerbestimmung der antilymphozytären und der antigranulozytären. Wirkung des Antilymphozyten-Immunserums (ALS)," in R. Pichlmayr and colleagues *Beitr. z. gerichtl. Medizin.* Franz Deuticke: Wien, 1970 (316-319).

Schulze, W. "Untersuchungen über die Kapillaren und postkapillären Venen lymphatischer Organe." *Z. Anat. Entwickl. Gesch.* 76 (1925): 421.

Schwamm, E. "Ultrarotbstrahlung und ihre medizinische Bedeutung." *Erfahrungsheilkunde.* 4 (1955): 481-501.

Seeger, P.G. "Vergleichende mikrochemische Untersuchungen an normalen Exsudatzellen und den Tumorzellen des Ehrlichschen Aszitskarzinoms der Maus. Der Kalium-Natriumkontrast." *mikr. anat. Forsch.* 53 (1943): 65-101.

Seelich, F. and L. Stockinger. "Ein Beitrag zum Problem der Zellform und deren Umwandlung." *Z. Zellforsch.* 39 (1953): 212-231.

Seeling, W. et al. "Die Funktion des Zinks im Organismus---dargestellt am Beispiel der Akrodermatitis enteropathica." *Med. Welt* 28 (NF 1977): 1537.

Sehrt, E. *Elektive Ultraviolettbestrahlung in Therapie und Prophylaxe.* Hippokrates-Verlag Marquardt u. Cie.: Stuttgart, 1942. Ch. IV and V include further bibliography regarding Havlicek.

Selye. H. *Einführung in die Lehre vom Adaptations syndrom.* G. Thieme: Stuttgart, 1953.

Sieberth, E. "Blutchemische Veränderungen im Verlauf der Karzinomkrankheit. Fortschritte der Krebsforschung." Report on 10. wissensch. Tag. des dtsch. Zentralausschusses für Krebsbek. und Krebsforschung e.V. Berlin, 1968, page 219 (includes bibliography).

Siegmund, H. "Speicherung durch Retikuloendotherlien, zelluläre Reaktion und Immunität." *Klin. Wschr.* 1 n. 52 (1922): 2566.

Sommer, H. "Erfahrungsbericht über Elpimed (früher Polyval)." *Med. Klinik* 48 (1952): 1224.

Speranksy, A. D. *Grundlagen der Theorie der Medizin.* Deutsch v. K P. von Roques. Verlag Dr. W. Sänger: Berlin, 1950.

Stacher, A. "Zur Wirkung der Herde auf den Gesamtorganismus." *Österr. Zeitschr. f. Stomat.* 63 (1966): 294-303.

_____. "Über das Huneke-Sekundenphänomen und seine Objektivierung," in Voss Ferd. *Deshalb Neuraltherapie.* Verl. Blume u. Co., 38-149.

Stern and Z. Willhelm. *Krebsforch.* 46 (1937): 379. Qtd. in Hinsburg, *Krebsproblem.*

Stockinger, L. "Disse'scher Raum und Bindegewebszellen in der menschlichen Leber." *Anat. Anz. Erg. H.* 120 (1967): 545-551.

_____. "Ultrastruktur und Histophysiologie der menschlichen Leber. (Wien)." *Klin. Wschr.* 81 n. 23 (1969): 431.

_____. "Morphologische Grundlagen der Erregungsübertragung in zellulären Systemen." *Wissenschaft und Weltbild* 25: 73-87.

Stockinger, L. and W. Pritz. "Morphologische Aspekte der Schmerzempfindung im Zahn." *Dtsch. Zahnärztl. Zschr.* 25 (1970): 357-363.

Stöhr, jr. Ph. "Mikroskopische Anatomie des vegetativen Nervensystems." *Handb. d. mikr. Anat. d. Menschen v. Muollendorf-Bargmann*, Bd. IV/5. Springer, 1967.

Storck, H. *Rheumatismus als Regulationskrankheit*. Verlag Urban und Schwarzenberg: München/Berlin, 1954.

Undritz, E. "Über das Vorkommen von Abbauformen von Leukozyten im Blut." *Fol. hämat.* 65 (1941): 195.

Undt, W., H. Karobath, and Mitarb. "Der Herzmuskelinfarkt. Untersuchungen über den Einfluß von Wetter etc. auf den Zeitpunkt des Krankheitsbeginnes." *Z. f. angew. Bäder- und Klimakunde* 19 (1972): 2-3.

Urbach, E. *Klinik und Therapie der allergischen Krankheiten*. Verlag W. Maudrich: Wien, 1935.

Virchow, R. *Die Cellularpathologie in ihrer Bedeutung auf physiologische und pathologische Gewebslehre*. Hrischwald: Berlin 1858.

Voss, H. F. "Deshalb Neuraltherapie; 70 Arbeiten über die Neuraltherapie." Ml-Verlag. Schriftenreihe des Zentralverbandes d. Ärzte für Naturaheilverfahren, edited by v. Dr. H. Haferkamp, Bd. 20.

Zabel, W. "Ganzheitsbehandlung der Geschwulsterkrankungen." *Hippokrates* 31 (1960): 751-760.

_____. . *Körpereigene Abwehr gegen Krebs*. Med. Liter. Verlag: Uelzen, 1964.

Zen, Shomo. "Studien über Monocytie durch Organextrakt, besonders über die Bedeutung des RES, welches diese Organe enthalten." *J. Chosen med. Assoc.* 30 n. 1 (1940). German summary 173-174.

Zischka, W. *Infektionskrankheiten*. Verlag Urban und Schwarzenberg: München/Berlin, 1961.

Zypen, E. van der. "Elektronenmikroskopische Befunde an der Endausbreitung des vegetativen Nervensystems und ihre Deutung." *Acta anat.* 67 (1967): 431-515.

Index

A

Acothiaprine, 174
Acupuncture points
   changes in physical functions of, and disease, 108–9
   correlates to, 39–40
   diagnosis and, 106–15
   effectiveness of, 182
   electrophysiological phenomena and, 110–15
   maximum points and, 96
   meridians and, 104
   morphology of, 106
   palpation of, 109–10
   rhythm harmonization and, 114
   stimulation of, 107–8, 114
   thermal phenomena and, 110
Adaptation syndromes, 84, 102
Afferents, spinal, 97–98
Aging, 18
Alarm reaction, 63, 126–27, 134, 136, 140, 150, 178
Alzheimer's disease, 18–19
Amygdala, 36
Amyloid, 18
Antibiotics, 174
Antioxidants, 16
Arteriosclerosis, 18, 35
Arthritis, rheumatoid, 157, 173–74, 183
Ascorbic acid, 16
ATP synthesis, 32
Aureotherapy, 174
Autoimmune diseases, 71
AVA (arterio-venous anastomoses), 103–4
Axial structure, role of, 97

B

Balneology, 182
Basement membranes, 46
Biocybernetics
   chronicity viewed as problem in, 86–102
   principles of, 84–85
Biological systems
   nonlinearity of, 7–8
   redundancy of, 14
Blood, oxygen saturation of, 132–33
Body awareness, 35–36
Bruno, Giordano, 72

C

Calcium deficiency, 177, 179
Cancer treatment, 166–71. *See also* Tumors
Capacity, measuring, 118
Carbohydrates, storage of, 18
Carcinoma. *See* Cancer treatment; Tumors
Cardiac crises, 166
Catalase, 16
Catecholamines, 36–37, 42
Cell
   as central point in connective tissue, 7
   as functional unit, 4–5
   relationship of, with environment, 4
   surface, sugars of, 43–45
Chemical sensitivity, 16–17
Chondroitin sulfate, 41
Chondroitin sulfate proteoglycan (CSPG)
   –hyaluronic acid complexes, 51–52
   structure of, 50–51
Chondronectin, 65
Chronic conditions
   as biocybernetic problem, 86–102

iceberg analogy for, 90, 98, 101
pathogenesis of, 90–102
recording immune reactions during, 135–50
time and, 98–99
Chronobiology, 32–33
Circadian rhythm, 32–33
Clinical trials, insufficiency of, 3–4
Collagen
  antigenicity of, 58
  formation of, 44
  functional aspects of, 58–59
  modification of, 57–58
  molecular and supramolecular structure of, 56–57
  protein stored as, 17
  synthesis of, 16, 55–56
Colloidal state, 103–4
Conductance
  investigations of, 110–12
  measuring, 117
Connective tissue
  cell as central point in, 7
  function of, 4–6
  loose, 39
Contact irradiation, 167
Coping process, 34
Corticoidsteroids, 174
Cybernetics, 7, 82–85. *See also* Biocybernetics
Cyclosporin, 174

D

Degenerative deterioration, 101
Deqi, 107–8
Dermatin sulfate proteoglycan (DSPG), 52–53
Diabetes, 18
Diagnosis
  acupuncture points and, 106–15
  criteria of, 103–4
  methods of, 116–18
  somatotypes and, 104–5
Digestive tract, disturbance of, 160–63, 174–75
Disease. *See also* Chronic conditions
  acupuncture points and, 108–9
  bodily fluids and, 7
  course of, 172
  extracellular matrix and, 6, 15, 16, 172
  modern medicine's model of, 3
  palpation reflex signs of, 116
Disturbance fields, 101–2, 146, 160, 163, 179–80
DNA, 11, 14, 15, 42
Drug selection, 173–75

E

Elastin
  functional aspects of, 60–62
  molecular and supramolecular structure of, 59–60
  synthesis of, 16, 44, 59
Electrodiagnosis, 117–18
Electromagnetic signals, measuring, 118
Electro skin test, 117
Endocrine gland system, 13
Energy flow, 66–69
Enthalpy, 67
Entropy, 67
Environmental medicine, 16–17
Epithelium, growth of, 46
Eutectic point, 48
Extracellular matrix. *See also* Matrix regulation; Matrix synthesis
  acupuncture points and, 106–15
  biochemistry of, 8
  definition of, 4
  disease and, 6, 16, 172
  distribution, topography of, 39–40
  encoding of information in, 8, 11, 14
  endocrine gland system and, 13

energy flow in, 66–69
evolutionary age of, 14
homeostasis and, 8
metabolism in, 11–16
organization of, 4, 48
as protein regulator, 17–19
redox potential of, 16, 48
structural components of, 41–42
toxins bound by, 16–17
tumor, 30–31
ultrastructure of, 10

F
Factor L, 155–56
Fibroblast, matrix-synthesizing, 10, 11
Fibrocytes, 14
Fibronectin, 62–64
Foci, 101–2, 163, 175–76, 180–81
Free radicals, 16, 17

G
Galen, 5
Galileo, 3
Gamma motor neuron, 94
Gastric acid deficiency, 178
Gate control system, 93–94
General adaptation syndrome, 33
Glutathione peroxidase, 16
Glycocalyx, 43–45
Glycogen, 18
Glycosaminoglycans (GAGs)
　charge of, 41
　formation of, 44
　glycocalyx and, 43
　important, 41–42
　in the matrisome, 11–12
　mesh of, 8, 10, 11
　protein stored as, 18
Goethe, Johann Wolfgang von, 72

H
Hahnemann, Samuel, 5, 68
HCl deficiency, 178
Heine cylinders, 99–100, 104
Hematogenous oxidation, 170–71
Heparan sulfate, 41
Heparan sulfate proteoglycan (HSPG), 53
Heparin, 41, 42
Hippocrates, 5
Histiocyte wall, 150, 151
Homeopathy
　effectiveness of, 182
　immunological bystander reaction of, 70–71
　nature of, 68–69
Humoral theory, 7, 13
Huneke, Ferdinand, 155, 160, 163, 164, 183
Hyaluronic acid, 11, 15, 33, 41–42, 51
Hyper-reactive allergy reactions, 87
Hypothalamus, 36–37, 97

I
Immune capacity, rehabilitation of, 175–76
Immune processes
　controlling, 151–59
　regulatory systems and, 172
Immunological bystander reaction, 70–71
Inflammation
　chronic, 101–2, 176, 180–81
　neurogenic, 37
Information
　encoding of, in the extracellular matrix, 8, 11, 14
　as energy carrier, 8
　transmission of, 99–100
Infrared diagnosis, 116–17
Intestinal flora, disturbance of, 160–63, 174–75
Iodometry, 125, 134
Iron deficiency, 179

K

Keratan sulfate, 41
Keratan sulfate proteoglycan (KSPG), 53–54

L

Laminin, 64–65
LDL cholesterol, 18–19
Lectins, 45
Leukocytes
   lysis of, 19–20, 21–30, 128, 130, 170–71
   observation of, for diagnosis, 105
Limbic system, 97–98

M

Macrophages, 14
Magnesium deficiency, 177, 179
Maladaptations, 17
Mast cells, 42
Matrisomes, 11–12
Matrix regulation
   diagram of, 9
   leukocytolysis and, 21–30
   maladaptation phase of, 17
   rhythms of, 32–33
   screening of, 17
   tumor, 30–31
Matrix synthesis, 10
Matrix vesicles
   disintegration of, 19–20
   tumor, 30
Maximum points, 96
Meridians, 104
Monocyte factor, 151–55
Multiple sclerosis (MS), 135–37, 157, 175–76, 183
Muscle activity, 105
Muscular maximum points, 96
Musculature, regulatory control of, 94–96
Myocardial infarction, 166

N

Neural therapy
   according to Huneke, 160–71
   for tension syndromes and tension pain syndromes, 88
Neurogenic inflammation, 37
Nucleotides, 15
Nutritional deficiencies, 177–79

O

Organism, as network system, 81
Orthodiagrams, 167
Overcompensation, 84
Oxygen, 15–16, 132–33

P

Palpation
   features of, 109–10
   reflex signs of illness, 116
Paracelsus, 5, 68
Parenchymal cells, 5
Perfusion, 104–5
Phenylbutazone, 173–74
Potential differences
   experiments of, 112–14
   measuring, 118
Projection symptoms, 104
Proteins
   crosslinked, 62–65
   regulation of, 17–19
   storage of, 17–18
Proteoglycans (PGs)
   chondroitin sulfate, 50–52
   dermatin sulfate, 52–53
   extracellular matrix homeostasis and, 8
   formation of, 44
   heparan sulfate, 53
   hyaluronic acid and, 15
   keratan sulfate, 53–54
   in the matrisome, 11–12
   mesh of, 8, 10, 11

molecular structure of, 47
nutritional groups stored by, 17–18
synthesis of, 16, 49–51
water molecules and, 47
Psychoneuroimmunology, 14
Puncture phenomenon
   bioelectric events during, 130–32
   description of, 125–30
   iodometry and, 125, 134
   leukocytolysis of, 128, 130
   oxygen saturation of blood and, 132–33
   regulation during, 134
   thermoregulatory phenomena and, 130–31

R
Radiation therapy, 167–69
Reflex pathway, viscerocutaneous, 91
Regulatory cycle
   diagram of, 82
   disturbance of, 83–85
   elements of, 82
   functioning of, 82–83
   as smallest cybernetic unit, 82
Regulatory disintegration, 101, 108
Regulatory therapy
   aim of, 119
   important facets of, 119
   for respiration and circulation, 89–90
   success and failure in, 182–84
Rehabilitation treatment, 181–82
Respiration, 89–90
Reticular fibers, formation of, 44
RNA, 11, 15, 42
Rokitansky, Carl von, 7

S
Scars, 164, 165, 179–80
Schelling, Friedrich, 72
Schizophrenia, 159
Segmental regulatory complex, 91–94, 101–2
Selye, Hans, 33, 85, 102, 126–27, 136, 140, 150, 154, 170
Sensorimotor control system, 91–98
Shock reactions, 154–55, 173
"Slag phenomenon," 17–19
Somatotypes, 104–5
Spherics, 99
Spinal cord, segmental arrangement of, 92
Spontaneity vs. chance, 67–68
Stress. *See also* Stress reaction process
   biological rhythms and, 33
   coping process for, 34
   definition of, 33
   factor, optimal, 35
   free radicals and, 17
   on immune regulatory cycles, 176–79
   minimal chronic, 101–2
   symptom, 96
Stress reaction process
   body awareness and, 35–36
   controllable and uncontrollable, 34–35
   diagram of, 38
   neurological basis of, 36–37
   phases of, 33
Structural glycoproteins, 11–12, 55–65
"Sugar principle of living," 15
Sugars
   of cell surface, 43–45
   as evidence of pre-cellular evolution, 72
Superoxide dismutase, 16
Surfaces, energetically minimal, 12
Synovial fluid, 11

T
Thermodiagnosis, 116–17
Tissue water, 47–48
Toxins, environmental, 16–17
Tuberculosis, chronic pulmonary, 86–90
Tumors
   extracellular matrix and, 30–31
   malignant, 157–59
Tunnel processes, 72

V
Virchow, Rudolf, 3, 5, 7
Vitamin A, 16
Vitamin E, 16

W
Water molecules, 47
White blood cells. *See* Leukocytes

X
X-rays, 167

Z
Zinc deficiency, 177, 178–79

About the Author

ALFRED PISCHINGER was born in 1899 in Linz, Austria. He earned his MD at the University of Graz in Austria in 1923. He continued his medical work at the Department of Histology and Embryology at Graz University, and became chairman and a professor of the department in 1933. In 1958, Pischinger became head of the Department of Histology and Embryology at the medical faculty of the University of Vienna, where he remained until his retirement in 1970. He died in 1983 in Graz.

Beginning in 1948, Pischinger was the first scientist to describe the regulation of the extracellular matrix (ground regulation) and to state that each disease starts within the extracellular matrix. This view led him to found the theory of the "ground regulation system" in the 1950s. This theory has developed as the scientific basis for complementary (holistic) medicine and has allowed for a better understanding of conventional medicine. In 1975, Pischinger presented his findings in the first edition of this book (published in German).

Since 1984 Pischinger's work has been continued by Hartmut Heine.

About the Editor

HARTMUT HEINE (born 1941 near Munich, Germany) is a former professor of anatomy and head of the Department of Anatomy and Clinical-Morphology of the University of Witten/Herdecke Germany. He studied anatomy and biology at the universities of Munich, Kiel, and Hannover, in Germany. In 1968, Heine earned a Doctor of Natural Sciences (*doctor rerum naturalium*) degree from the University of Kiel, and later earned his PhD in Anatomy from Frankfurt/Main University. In 1976, Heine was appointed Professor of Anatomy at the University of Würzburg Germany, and 1982 he become full professor at Witten/Herdecke University.

Heine's main scientific work deals with the comparative anatomy of the vertebrate heart, microcirculation, and the structure and function of the extracellular matrix. He lives near Stuttgart, Germany.

About the Translator

INGEBORG EIBL was born in Germany and moved to the US as a child. She studied music and pre-med at McGill University, and later graduated from New York Chiropractic College and Ontario College of Naturopathic Medicine. She studied homeopathy and gentle effective manipulative therapies. Eibl has a chiropractic practice in Rochester, New York.